SPACE

SPACE
our final frontier

John Gribbin

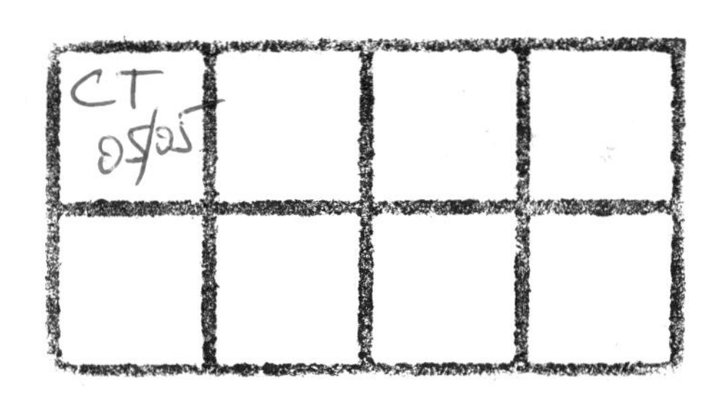

This book was originally published to accompany the BBC television series *Space*, which was first broadcast in 2001.

Creative Director: John Lynch
Executive producer: Emma Swain
Series producer: Richard Burke-Ward

Published by BBC Worldwide Limited,
Woodlands, 80 Wood Lane, London W12 0TT

Hardback edition first published 2001
Reprinted 2001
This paperback edition published 2003

ISBN: 0 563 48893 X

Commissioning Editor: Joanne Osborn
Project Editors: Helena Caldon and Julie Tochel
Production Controller: Christopher Tinker
Academic Consultant: Dr Margaret Penston
Design concept: The Attik
Art Direction: Pene Parker, Lisa Pettibone
Art Editor: Kathy Gammon
Design: Bobby Birchall, DW Design, London
Picture Research: Carmen Jones and Miriam Hyman
Illustrations: Mark McLellan and Kevin Jones Associates

Set in Albertina MT and Gill Sans
Printed and bound in Great Britain by Butler & Tanner Ltd, Frome
Colour separations by Radstock Reproductions Ltd, Midsomer Norton

1

ACROSS THE UNIVERSE

1.1 STEPPING STONES TO THE UNIVERSE

The travellers in *Star Trek* boldly go where no one has gone before to explore the final frontier – space. No human has yet visited any object outside our own Solar System, but that has not stopped us exploring new worlds – at long range, using telescopes on the surface of the Earth and satellite-based instruments orbiting above the atmosphere. The data from these observations are then compared with what we can infer about stars and galaxies from the laws of physics. In astronomy, theory and observation always go hand in hand – a theory about stars is useless without observations to test the predictions of the theory, and observations of a startling new phenomenon remain a mystery until they can be understood within the frameworks of a theory about the Universe. Together, theory and observation can take us on a journey to the furthest reaches of the Universe, and back in time to when the Universe was born.

Previous page. A spiral galaxy, like the Milky Way in which we live.

MAKING MAPS OF SPACE

Astronomers are interested in the evolution of stars and galaxies (how these objects are born, live and die), and in tracing the origin and ultimate fate of the entire Universe. Putting this type of knowledge together with a knowledge of the distances between cosmic objects enables astronomers to achieve an understanding of their domain similar to a naturalist's understanding of the world – the equivalent of combining biology with geography, and finding out what different kinds of creatures live in different parts of the world.

By studying the light emitted by stars and galaxies, astronomers are able to find out what different kinds of objects exist in different parts of the Universe. But they also need to measure the distances to cosmic objects, so they know where they are in relation to one another. And how can the distances to stars and galaxies that we have no hope of ever visiting, even with an unmanned space probe, be measured? It sounds like an impossible task, but astronomers have found 'stepping stones' that effectively take them from Earth to the furthest reaches of the Universe.

It's all Done with Triangles

As the Chinese proverb says, the longest journey begins with a single step; and the geographical exploration of the Universe starts with a simple piece of geometry involving triangles.

The first step into the Universe uses exactly the same kind of surveying techniques used here on Earth to measure the distances to distant objects (such as mountains) without actually having to go there. The idea itself is not new, but with the aid of new instruments here on Earth and satellites orbiting above the Earth it reaches further than ever before.

It all depends on the geometry of triangles. If you know the length of one side of a triangle (the base) and you can measure the angle each of the other two sides makes with the base, then it is a simple matter to calculate how far it is from the base of the triangle to

1. Human beings have only visited our nearest neighbour in space, the Moon.

2. Exploration of deep space relies on remote observing using instruments such as radio telescopes.

the opposite tip. The process is called triangulation, for obvious reasons.

The trouble with triangulation is that you need a longer baseline to measure the distances to more distant objects.

The Importance of Parallax

Triangulation is not restricted to measuring distances on Earth – it works very well for measuring the distance to our nearest neighbour in space, the Moon. If one observer sees the Moon directly overhead, for example, while another observer, standing on what is the horizon for the first observer, also measures the angle it is in the sky, then it is easy to work out the distance to the Moon – about 384,000 km – from the geometry of triangles.

This is possible because the Moon appears in a different part of the sky to the two observers. It is exactly the same as the way you can make your index finger jump across a distant background if you hold your arm straight out in front of you, point your finger upwards, and look at it with each eye closed in turn. The slightly different views from your two eyes give you different perspectives on your finger; and the slightly different views from two observatories give different perspectives on the Moon. The effect is called 'parallax', and for our two observers it shifts the apparent position of the Moon nearly twice the diameter of the full Moon on the night sky.

But the distant stars are so far away that the background pattern looks the same from anywhere on Earth, and this makes it convenient to measure parallax by measuring how far the Moon (in this case) seems to shift against the fixed background of stars.

1

MAPPING THE EARTH

Triangulation is a technique that can be used to measure distances to distant objects. A map-maker can plot the location of a hill on a map by first measuring out a baseline, perhaps a kilometre or so long then, using small telescopes (called theodolites), measuring the angle between the baseline and the hilltop – from both ends of the baseline. Using these angles, and knowing the length of the baseline, the map-maker can now calculate the lengths of the other two sides of the triangle, and thus how far the hill is from the baseline. The distance from one end of the first baseline to the hilltop can be used as a new baseline to measure other distances. You can use this process repeatedly to work your way across the surface of the Earth. Indeed, although the technique has now been superseded by satellite mapping, triangulation was used by nineteenth-century surveyors to map India, starting at the southern tip and working their way northward to the Himalayas.

Beyond the Moon

Triangulation and parallax have also been used to measure the distances to the nearest planets, Venus and Mars. This is much harder than measuring the distance to the Moon, because the planets are much further away. It involves making observations from opposite sides of the Earth at the same time, then calculating the geometry of a very tall, thin triangle.

The parallax of Mars was determined accurately in 1671, when the French astronomer Jean Richer led an expedition to French Guyana (in South America) to measure the position of Mars against the background stars at a certain time on an appointed night (actually several nights, to allow for cloud).

On the same nights and at the same times, back in Paris, the Italian-born astronomer Giovanni Cassini also made observations of the position of Mars against the background stars. When Richer's expedition returned, the two teams compared notes and calculated the distance to Mars.

Law-abiding Planets

These measurements were particularly important, because they made it possible to work out the geography of the entire Solar System.

The laws which describe the motion of the planets around the Sun were described early in the seventeenth century by Johannes Kepler, and explained by Isaac Newton with his theory of gravity. They state that, if planet A is twice as far from the Sun as planet B, then the orbital period of planet A (the time it takes to go round the sun: its 'year') is a certain multiple of the orbital period of planet B.

Astronomers thus had to measure at least one planetary distance directly in order to put real numbers into the equations, even though they already knew the orbital periods for the planets. By measuring the distance to just

The observations that led to the first measurement of the distance to Mars were made in French Guyana, near the site used by the European Space Agency to launch its Ariane rockets.

2

1. Triangulation and the parallax effect.

2. Once astronomers had determined the sizes of galaxies, they could even use triangulation to estimate the distances between them from how small the galaxies look on the sky.

1 and 3. Johannes Kepler discovered the laws of planetary motion by studying the orbit of Mars (right).

2. Kepler's key discovery was that planets in their orbits trace out equal areas in equal times.

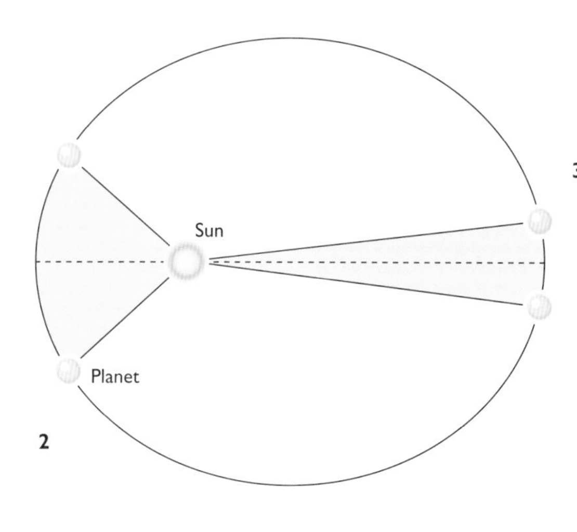

Venus and Mars, they were able to calculate the distance from the Sun to each of these planets. Once they knew these distances, they were able to use Kepler's laws to calculate the distance from the Sun to all the other planets in the Solar System – including the Earth. In addition, they could use Newton's laws to calculate what the mass of the Sun must be to hold the planets in these orbits through gravity.

By the end of the seventeenth century, astronomers were able to calculate the distance from the Earth to the Sun fairly accurately. The observations have been improved since then (we can even measure the distance to Venus directly, by bouncing radar signals off it), and the distance from the Earth to the Sun is now known to be 149.6 million kilometres (almost 4000 times the distance around the equator of the Earth). But even 200 years ago, the calculated distance was 140 million km – an error of less than 7 per cent compared with the modern figure.

STEPPING STONES TO THE STARS

It takes the Earth 12 months to orbit the Sun once. The radius of the Earth's orbit – the distance from the Earth to the Sun – is roughly 150 million km. This distance is called the Astronomical Unit, or AU. It is vitally important in astronomy, because it provides a new baseline with which to measure parallaxes for more distant objects – the nearest stars.

At intervals 6 months apart, the Earth is at opposite sides of a diameter measuring 2 AU (about 300 million km). This is such a long baseline that in photographs of the night sky taken 6 months apart a few of the stars seem to have shifted their position slightly, because of the parallax effect. But the shift is very slight, because the stars are so very far away. To give you some idea how small the effect is, in the 1830s the first star studied in this way (known as 61 Cygni) was found to have a parallax shift of just 0.31 seconds of arc. (There are 360 degrees in a circle, 60 minutes in a degree, and 60 seconds in a minute.) In comparison, the full Moon covers 30 seconds of arc on the sky. So the apparent shift in 61

Mount Everest is named after Sir George Everest, the surveyor who led the team that mapped India in the nineteenth century.

4. We live in a spiral galaxy like this one (M65) containing hundreds of billions of stars like the Sun.

1

Cygni as the Earth goes round the Sun is equivalent to about one-six-thousandth of the apparent diameter of the Moon.

The distances to the stars are so great that astronomers had to invent new units with which to describe them. If you were so far away from Earth that the distance between the Earth and the Sun (the radius of the Earth's orbit, 1 AU) covered just one second of arc on the sky, then you would be one parsec away from Earth (the term 'parsec' is a contraction of 'parallax second of arc'). A parsec is just over 30 million million kilometres, a number so big that it is hard to visualize – but you can look at it in terms of the speed of light. Light travels at just under 300,000 km per second, and so covers 9.46 million million km in a year, a distance known as a light year. So a parsec is 3.26 light years. Converting the parallax measurement into distance, we find that 61 Cygni is 3.4 parsecs away, or just over 11 light years from us. And, amazingly, this makes it one of the closest stars to our Sun.

Stars Like Dust

When you look up at the sky on a dark and cloud-free night, it seems to contain countless numbers of stars, and poets have waxed lyrical about the view. But the human eye is not very sensitive to faint light: even under perfect conditions, with no Moon or cloud, and far from city lights, the most you can see at any one time is about 3000 stars. Under more ordinary viewing conditions, you are lucky to see a thousand.

The true numbers of stars in the sky only began to be appreciated at the beginning of the seventeenth century, when Galileo Galilei turned his telescope onto the night sky. He found that, what seemed to be a faintly glowing cloud of light was actually a myriad of individual stars, each too faint to be seen by the unaided human eye. He announced his discoveries in a book, *The Starry Messenger*, which was published in 1610.

At that time there was no accurate way to estimate the distances to the vast majority of these stars. Until very recently, only a few stellar distances had been measured directly by parallax. By the end of the nineteenth century, just 60 stellar distances had been measured in this way. At the end of the twentieth century, the situation improved dramatically when the HIPPARCOS satellite, orbiting clear of the obscuring influence of the Earth's atmosphere, measured the distances to a large number of stars with unprecedented accuracy. It pinned down the parallaxes of more than 100,000 stars, to an accuracy of 0.002 seconds of arc. But even this impressive achievement gives the distances to less than one-millionth of the total number of stars in the Milky Way, taking the range of directly measured stellar distances out to a few hundred parsecs (about a thousand light years).

Colour, Brightness and Distance

So even with the aid of satellites like HIPPARCOS, astronomers still need other techniques to measure distances to stars outside our local region of space. The most important of these techniques is called the 'moving cluster' method. It gives the distance to a large group of stars, called the Hyades Cluster. These stars are about 40 parsecs (130 light years) away from us, and all move as a group through space (HIPPARCOS has, now, also confirmed the distance to this cluster). Because this cluster contains hundreds of stars with different colours and brightnesses, the fact that they are all at the same distance away helps astronomers to understand how brightness is related to the colour of the light emitted, in subtle ways. The colour of a star has no relation to its distance: it's the brightness that tells us how far away it is. Then, when they

2

1. The satellite HIPPARCOS, shown here being tested prior to launch, measured the distances to the stars with unprecedented accuracy.

2. The Hyades star cluster and the smaller cluster known as the Pleiades.

THE STARS ARE SUNS

The Sun is a star or, to put it another way, the stars are suns. Scientists only began to appreciate this in the seventeenth century, and even then they found it hard to believe. Isaac Newton, for example, calculated that if the star Sirius is really as bright as our Sun, then to appear as faint as it does in the sky it must be a million times farther away from us than the Sun. Newton was more or less right, but he was so discomfited by this calculation that he never published it during his lifetime (it appeared in a book, *System of the World*, published in 1728, the year after Newton died). Yet Sirius is actually one of our nearest stellar neighbours, only 2.67 parsecs (8.7 light years) away.

We now know that the Sun is a very ordinary star. It is neither particularly large nor particularly small, not unusually hot nor unusually cool, and it is roughly halfway through its life. This is disappointing for people who like to think that there is something special about our place in the Universe. But it raises a much more exciting possibility. If the Sun is an ordinary star, then it seems likely that other stars like the Sun will have families of planets orbiting them like the Solar System. Could it be that as well as being ordinary in every other way, the Sun is quite typical in having a family of planets that includes a home suitable for life? If so, there may be literally billions of other Earths out there in the Milky Way Galaxy.

SPECTROSCOPY: THE KEY TO ASTRONOMY

The single most important tool of astronomers is the ability to analyze starlight and discover what stars are made of. This depends on the fact that atoms of any particular chemical element radiate energy (if they are hot) or absorb energy (if they are cold) at very precise wavelengths in the rainbow spectrum of light. Each element produces its own distinctive set of lines in the spectrum when radiating or absorbing energy, yielding a pattern similar to a barcode. And, like a barcode, each pattern is unique.

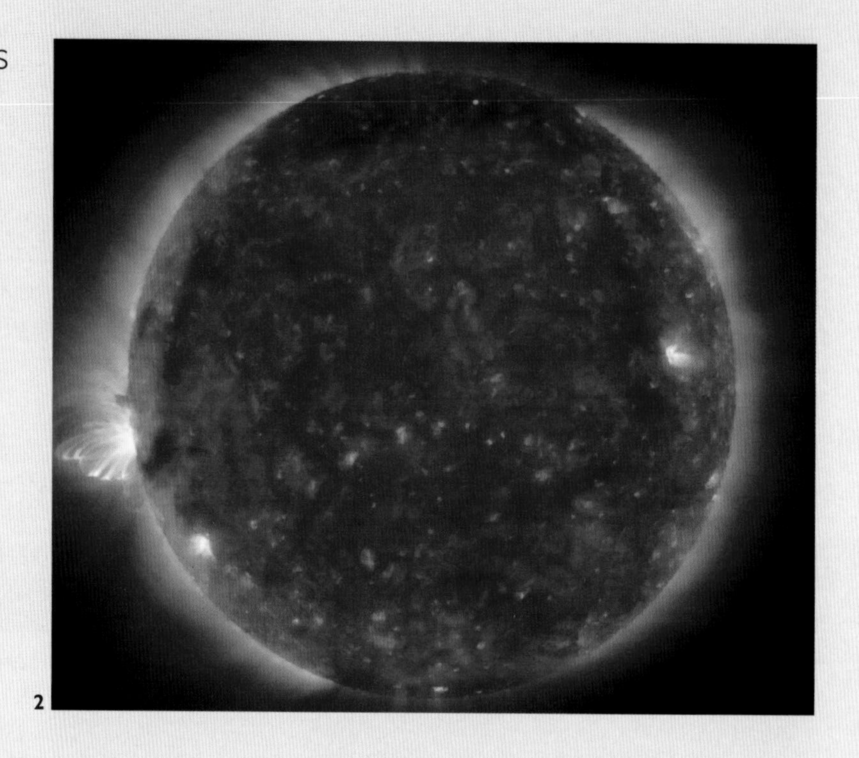

2

1

The Flame Test

We know which spectroscopic 'barcode' corresponds to a particular element because the light emitted by elements has been studied using simple flame tests. A sample of a known element (perhaps a piece of copper wire) is heated (often using a simple Bunsen burner) and the light it radiates when it is heated is passed through a triangular prism. This spreads the light out and produces a pattern of lines that is unique to that element. By repeating this test with many substances, a huge library of patterns has been built up, and an unknown substance can be identified by examining its spectroscopic pattern and comparing it with the patterns in the library.

Spectroscopy was invented in the middle of the nineteenth century, but there were many gaps in this library of knowledge, which still needed much investigation to fill. The first person to notice that light from the Sun, when passed through a prism to make a spectrum, contained many distinct lines was the British physicist William Wollaston, in 1802. But he had no idea what they were. In 1814, the German Josef von Fraunhofer counted 574 lines in the spectrum produced by light from the Sun, and discovered many of the same lines in light from the stars. But the person who explained that these lines were caused by the presence of different elements in the atmospheres of stars was Gustav Kirchoff. He pioneered the basic principles of scientific spectroscopy in collaboration with Robert Bunsen in Germany at the end of the 1850s.

Secrets of Sunlight

Spectroscopic studies of the light from the Sun's atmosphere, obtained during an eclipse in 1868, showed a distinctive pattern of lines which did not correspond to any known element. The British astronomer Norman Lockyer concluded that there must be an element in the Sun that had never been found on Earth, and gave it the name helium, from helios, the Greek word for the Sun. Helium was actually identified on Earth in 1895, and Lockyer received a knighthood (partly as a result of his famous prediction) in 1897. Spectroscopy had actually found an element in our nearest star before it had been found on Earth.

Moving Stars

There is one other vitally important use of spectroscopy in astronomy. Although the lines corresponding to a particular element are always produced at the same distinctive wavelengths, if the object making the lines is moving, the whole barcode pattern is shifted across the spectrum. If the object is moving towards us, the lines are shifted to shorter wavelengths. Because blue light has shorter wavelengths than red light, this is called a 'blueshift'. Similarly, if the object is moving away from us, the pattern is shifted towards longer wavelengths, a 'redshift'. This is known as the Doppler effect, and it enables astronomers to measure how fast stars are moving through space, how fast galaxies are rotating, and how fast stars in binary systems (where two stars orbit around each other) are moving in their orbits. The last application is particularly useful, because it is a key (along with the law of gravity) to measuring the masses of the stars involved.

So spectroscopy tells us what stars are made of, how fast they are moving, and what mass they have. Without spectroscopy, there would be little more to astronomy than making pretty patterns called constellations out of the arrangement of stars on the night sky.

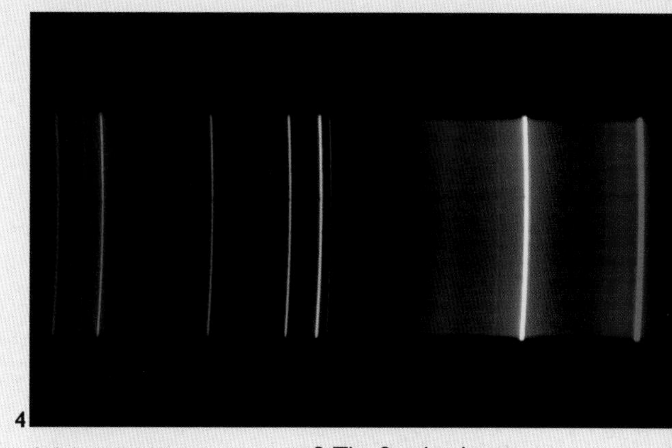

4

3

1. A bunsen burner being used to heat copper wire in a flame test

2. Solar corona - an ultraviolet image of the sun's outer atmosphere taken by the SOHO satellite. On the left is a prominence being absorbed after a solar flare event.

3. The Stephen's Quartet group of galaxies and NGC 7320. Colour coding shows the different red shift values of the quartet members.

4. The distinctive fingerprint of coloured lines is clearly shown here in the light from hot Helium gas.

TOPIC LINKS

1.1 Stepping Stones to the Universe
p. 17 Colour, Brightness and Distance

2.1 The Big Bang
p. 81 Galaxies on the Move

3.1 Life and the Universe
p. 140 The Big Surprise
p. 150 Seeking Signs of Life

1

2

see a star with the same colour as one of the types of Hyades stars, they can estimate its distance by comparing its brightness (or faintness) with the Hyades star. Crucially, the subtle differences in the colour of the stars involved are revealed by a technique called spectroscopy – probably the single most important tool used by astronomers.

The result of applying such methods is that we now have a clear idea of distances between stars, and also of their sizes. The distance from one star to even its nearest neighbours is usually tens of millions of times its own diameter (except, of course, for systems where two or more stars orbit around one another). For example, the Sun has a diameter of 1.39 million km (which is typical for a star during the main period of its life). If the Sun were the size of an aspirin, on this scale the nearest star would be another aspirin 140 km away. The distances between stars are absolutely enormous – even compared with the huge sizes of the stars themselves.

AN ISLAND IN SPACE

By using every possible technique for measuring distances to stars, astronomers have been able to map the collection of stars in which we live, an island in space called the Milky Way Galaxy, or just the Galaxy. This is rather like trying to map a forest from the inside by working out the distances and relative positions of the trees in all different directions. The process is aided by the fact that, in parts of the Galaxy, there are great clouds of gas and dust between the stars. These clouds contain large amounts of hydrogen, which can be detected by radio telescopes.

The overall shape of our Galaxy is a flattened disc containing hundreds of billions of stars, all more or less the same as our Sun, with a diameter of about 28 thousand parsecs (28 kiloparsecs). The disc is only 300 parsecs thick at its outer regions (roughly 1 per cent as thick as its width), but it has a bulge in the middle measuring 7 kiloparsecs across and 1 kiloparsec thick. If we could view our Galaxy from the outside it would look rather like a huge fried egg.

Surrounding the whole disc is a halo of about 150 known bright star systems called globular clusters. Each globular cluster is a ball of stars containing hundreds of thousands, or even millions, of individual

★ If you take a deep breath, you will have more molecules of air in your lungs than there are stars in all the galaxies in the visible universe put together.

3

1. An impression of the number of stars in our galaxy is given by this picture of part of the Milky Way.

2. The Milky Way has been mapped using star clusters.

3. Our Milky Way Galaxy is a flattened disc of stars embedded in a halo of globular clusters.

1. The Hooker telescope – used in working out the size of our galaxy.

stars, so close to one another that there may be 1000 stars in a single cubic parsec of space. From the way stars move, astronomers also infer that there is a great deal of dark matter (material that has mass but cannot be detected directly ▷ p. 96) surrounding the whole Galaxy and holding it in a gravitational grip.

Stars in Spiral Patterns

Viewed from above, our Galaxy has a distinctive structure, with bright trails of stars, called 'spiral arms', twining outwards from the central bulge. This is a very common feature of disc galaxies like the Milky Way, so much so that they are sometimes referred to as spiral galaxies. The most important distinction between the central bulge and the disc proper, however, is that the stars in the bulge (and the globular cluster stars in the halo surrounding the Galaxy) are all old stars. They are perhaps 12 billion years old and are known, for historical reasons, as Population II stars. There is also very little gas or dust in the bulge. The disc, where the spiral arms twine outwards, contains gas and dust and some old stars, but also middle-aged stars and very young ones, which are known as Population I stars (the Sun is a Population I star). New stars are still being formed in the disc all the time.

All the stars in the disc, together with the gas and dust, orbit around the centre of the Galaxy. But the disc does not rotate as if it were a solid object (the way a CD rotates as it is played). Each star moves independently – just as each planet in our Solar System orbits around the Sun independently – and the stars closer to the centre move faster than those near the edge of the disc. The Sun is travelling at a speed of about 250 km per second in its own orbit around the centre of the disc, carrying our Solar System with it; but the Galaxy is so large that even at this speed it takes our Solar System about 225 million years to complete just one orbit, a journey it has made about 20 times since it was born some 4.5 billion years ago.

The Sun and its family of planets orbits the Galaxy at a distance of about 9 kiloparsecs from the centre, two-thirds of the way out to the edge of the disc, on the inside edge of a feature known as the Orion Arm. We are not in the centre; there is nothing particularly special about our place in the Milky Way Galaxy.

A MATTER OF PERSPECTIVE

The size and shape of the Milky Way Galaxy were only really described properly in the 1920s. Before then, most people thought that the stars they could see in the sky made up the entire Universe – everything there was to see. But as well as revealing a myriad of stars in the Milky Way, telescopes also showed up faint patches of light in the sky, fuzzy blobs called 'nebulae'. At the same time that astronomers started to appreciate and understand the geography of the Milky Way, some of them began to wonder whether these nebulae might be other islands in space, galaxies like the Milky Way, but so far away from us that the light from the stars they contained only added up to a faint patch of light like a little cloud on the night sky. This suggestion caused a fierce debate among astronomers at the time, because it would mean that the other galaxies were at enormous distances from us, hundreds or even thousands of kiloparsecs away. This was hard to accept when astronomers had only just discovered that the Milky Way itself was several tens of kiloparsecs across, bigger than anything previously imagined. The other possibility was that the nebulae were indeed glowing clouds of gas within the Milky Way itself, between the stars.

The only way to find out whether the nebulae were actually galaxies was to identify individual stars within them and measure their distances directly. They would be too far away for triangulation to work; but by the 1920s astronomers knew that some kinds of exploding star (called novae) all have about the same brightness, while another kind of star (called 'Cepheids') have a brightness that can be inferred from their other properties. If you know the true brightness of a star, it is easy to work out how far away it is by measuring how bright it appears. So if the astronomers could identify novae and Cepheids in nebulae, they would be able to work out roughly how far away they were.

2

2. Supernova 1987A, a huge stellar explosion photographed in March 1987.

If each star were represented by a single grain of rice, a scale model of the Milky Way galaxy would just fit into the gap between the earth and the Moon.

It turned out to be just possible to make the crucial measurements for stars in some of these nebulae in the 1920s, using what was then the best telescope in the world. The telescope (still in use today) has a 100-inch diameter mirror, and is called the Hooker Telescope, after the benefactor who paid for it. It is located on top of Mount Wilson, near Pasadena, in California.

Beyond the Milky Way

The astronomer who made the crucial measurements was Edwin Hubble. He identified both Cepheids and novae in nebulae that are now known to be the closest galaxies to the Milky Way.

But it turned out that not all of the nebulae were other galaxies. Some of them really were clouds of gas and dust within the Milky Way, and these objects play an important part in the life cycles of stars and the origin of planetary systems like the Solar System.

In order to avoid confusion, astronomers kept the name nebulae for the clouds within the Milky Way, and used the term 'galaxy' to refer to the great star systems beyond the Milky Way.

Even with the 100-inch Hooker telescope, it was very difficult to make the observations needed to calculate the distances to galaxies. When Hubble first began to make

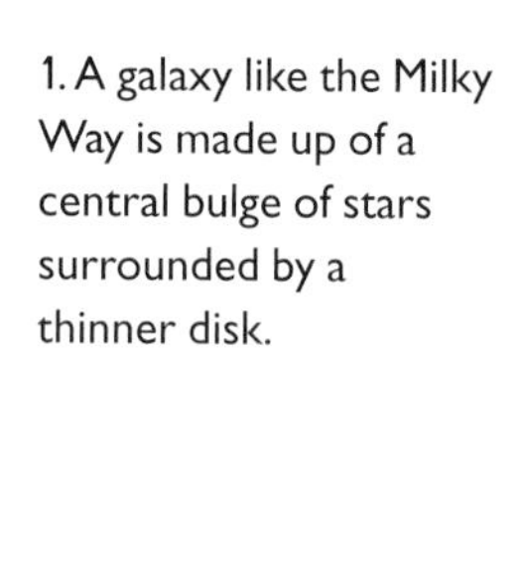

1. A galaxy like the Milky Way is made up of a central bulge of stars surrounded by a thinner disk.

2. (opposite) The Sun and the Moon both look the same size from the Earth even though the Sun is really much bigger than the Moon. In the same way, galaxies look tiny on the sky because they are so far away.

measurements of the distances involved, he found that, although the other galaxies did indeed lie beyond the Milky Way, they did not seem to be as big as ours. It's all a matter of perspective. One of the few things we can actually measure is the area a galaxy covers on the sky. A small galaxy close up will cover the same area as a big galaxy further away, in just the same way that if you hold a small model of a cow up in front of your eye, it looks as big as a real cow on the other side of a field. In the same way, the Moon completely covers the Sun during a total solar eclipse because, although the Sun is almost 400 times bigger than the Moon, it is also almost 400 times further away and so it appears to be the same size.

As telescopes got better and better, astronomers were able to measure the distances to other galaxies more and more accurately. They used many different stepping stones, not only Cepheids and novae, but also comparisons of the brightness of things such as globular clusters in one galaxy with those in another. After more than half a century of effort, they found that the galaxies were about 10 times further away than Hubble had thought, so it followed that they must be that much bigger than he had thought in order to look as large as they did on the sky.

However, Cepheids and novae are still the best indicators of distance. In the 1990s, using Cepheid distances obtained from the Hubble Space Telescope, a team at the University of Sussex finally showed that the Milky Way Galaxy is an average galaxy of its type (if anything it's a little bit smaller than the average disc galaxy in our part of the Universe). Like our position in it, there is nothing special about the Milky Way Galaxy.

Putting Galaxies in Perspective

The result of all these efforts is a clear understanding of the sizes of galaxies and the distances between them.

As well as disc (spiral) galaxies like the Milky Way, there are much larger, elliptical galaxies, which do not have a disc or spiral shape, but are ellipsoidal (like a rugby ball). These are thought to have been built up by a kind of cosmic cannibalism, from mergers between disc galaxies.

There are also smaller elliptical galaxies (resembling the globular clusters ▷ pp. 21-2) and small irregular galaxies which have no distinct shape. The largest elliptical galaxies contain several thousand billion stars. Disc galaxies, such as the Milky Way, have diameters of a few tens of kiloparsecs and contain a few hundred billion stars.

Galaxies are much closer together, relative to their own size, than the stars are to one another. Again, it's a matter of perspective. If we adapt the aspirin analogy to galaxies, and represent the Milky Way by a single aspirin, we find that the nearest large disc galaxy to us, the Andromeda Galaxy, would be represented by another aspirin just 13 centimetres away. And only 3 metres away we would find a huge collection of about 2000 aspirins, spread over the volume of a basketball, representing a group of galaxies known as the Virgo Cluster. On a scale where a single aspirin represents the Milky Way Galaxy, the entire observable Universe would be only a kilometre across, and would contain hundreds of billions of aspirin. In terms of galaxies, the Universe is a crowded place.

1. Most galaxies are grouped together in associations that are called clusters. This is the centre of the Virgo cluster of galaxies, which is a crucial stepping-stone into the Universe at large.

THE CEPHEID SCALE

Variable stars called Cepheids are the key to measuring distances across the Universe. Each Cepheid goes through a very regular periodic change in brightness; some have periods as short as a day, some take 50 days or so, others run through their cycle between these extremes. While they are interesting in themselves, their importance in astronomy derives from the fact that the Cepheids provide a vital stepping stone to the Universe – they give us accurate distances to the nearest galaxies.

2

Cepheids are large, yellow stars which have an intrinsic brightness between 300 and 26,000 times that of the Sun. They are typically between 14 and 200 times the size of the Sun. The change in brightness of a Cepheid is associated with a rhythmic pulsation, as if it is breathing in and out. A Cepheid is faintest when its outer layer has expanded to its maximum size and has cooled down, and brightest when the outer layer has contracted to its minimum size and is hotter.

First Steps

In the early years of the twentieth century, Henrietta Swan Leavitt was working at Harvard College Observatory, studying Cepheids in a star system called the Small Magellanic Cloud (which we now know to be a small irregular galaxy in orbit around the Milky Way). She noticed that the brighter a Cepheid is, the more slowly it goes through its cycle.

This showed up clearly for Cepheids in the Small Magellanic Cloud because the cloud is so far away from us that the distance from one side of the cloud to the other is such a small fraction of the distance to the cloud, that all the stars in it can be regarded as being at the same distance from us.

So if one Cepheid star looks twice as bright as another, it really *is* twice as bright, not just closer to us.

Stellar Street Lights

Leavitt's discovery of the relationship between brightness and period for Cepheids meant that they could be used to measure relative distances within the Milky Way. The period–luminosity relationship might tell you that one Cepheid is intrinsically twice as bright (say) as another, and comparing their apparent brightnesses on the sky would then reveal their relative distances from us. But in order to use this information to measure actual distances across the Milky Way, the distances to at least a few Cepheids had to be measured directly.

1

At first, this proved possible for just a handful of Cepheids; but crucially there were stars in the Hyades Cluster, whose distance could be measured, that are similar to stars in clusters that contain Cepheids. These provided the calibration, so that the distance to any Cepheid could be worked out simply by measuring its period, calculating its intrinsic brightness, and comparing that with its apparent brightness. The technique only just worked, even as late as the 1980s, because there were only 18 Cepheids with reliably determined distances. However, data from the HIPPARCOS satellite, which became available

3

1. Henrietta Swan Leavitt, who discovered the value of Cepheid stars as distance indicators.

2. The small Magellanic Cloud lies 20,000 light years away from us and contains many Cepheid variables.

3. Everybody who lives in the Northern Hemisphere has seen at least one Cepheid variable – the bright star in this picture.

4. If we know how bright a star really is we can work out its distance from us by judging how faint it looks – just like street lights.

in the 1990s, provided direct distance measurements to some Cepheids, and improved the accuracy of the distance measurement for the Hyades Cluster, so that the Cepheid distance scale is more solidly based now than ever before.

Into the Universe

Other objects within the same galaxies (most notably supernova explosions) can now be calibrated for brightness against the Cepheid distance scale. Because supernovae are so bright, they can be used as distance indicators in galaxies far, far away.

4

TOPIC LINKS

1.1 Stepping Stones to the Universe
p. 24 Beyond the Milky Way

2.1 The Big Bang
p. 80 Beyond the Local Gro

2.2 Cosmology for Beginners
p. 95 Einstein's geometry

1.2 STAR BIRTH

Stars are not born in isolation, but in nurseries that may contain thousands or even millions of stars forming together. No astronomer has been able to watch a single individual star being born, but because we see stars in all stages of the process, it is possible to work out how it happens. In the same way, an alien visitor to Earth could work out the life cycle of a human being without waiting to watch an individual person being born, living and dying, but by studying a large group of people where all stages of the life cycle are present.

Fortunately for astronomy, our Solar System is located in a densely populated part of the Galaxy, close to a stellar nursery where new stars are being born all the time. This is the Orion Nebula. The nebula is visible through a small telescope or binoculars as a faint patch of light just below the belt of the constellation Orion, the Hunter.

no real surprise, since it is very unlikely that any cloud of gas will just hang around in space without rotating. When the cloud began to collapse, its rotation would have got faster, in exactly the same way that spinning ice skaters rotate faster by pulling their arms in to their sides. This is a feature of something called angular momentum, which all rotating objects have. The amount of angular momentum depends on the mass of the object, how this is distributed, and how fast the object is rotating. If most of the mass is far out from the centre, it will have more angular momentum than if it is concentrated near the centre; and if the object is rotating fast, it will have more than if it rotates slowly. So when a spinning object shrinks in size, it spins faster to keep the same angular momentum.

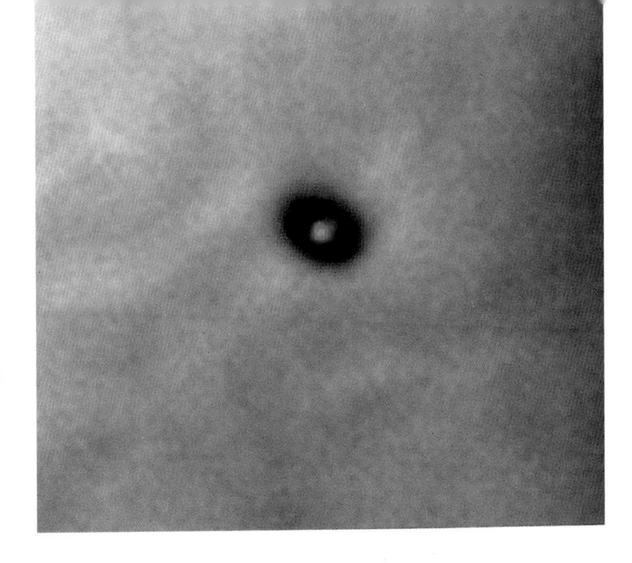

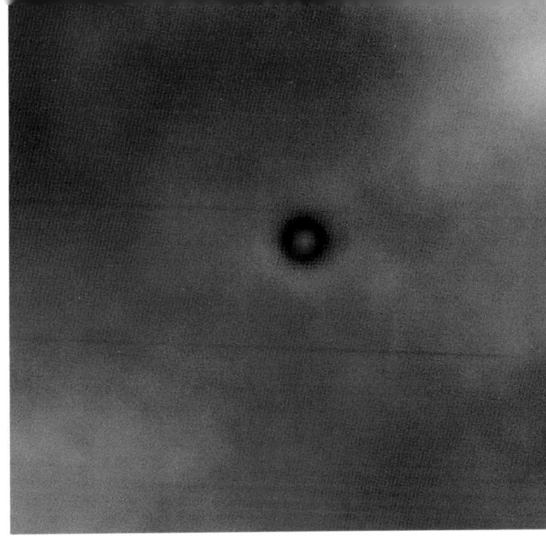

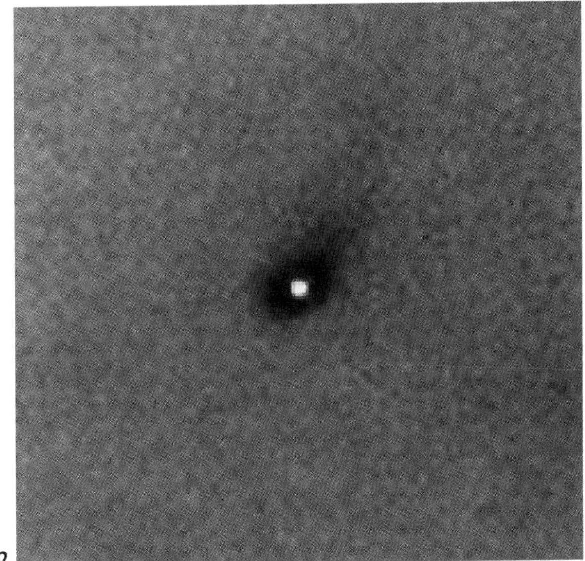

2

2. Disks of dusty material (protoplanetary disks) photographed around young stars in the Orion nebula.

In a Spin

The collapsing cloud that formed the Sun and planets got rid of some of its angular momentum by throwing material off into space, like water flung from a rotating garden sprinkler. But even so, there was too much angular momentum left to allow all the dust associated with the young star to fall into it. As the star began to shine at the heart of the collapsing cloud, the leftover dust settled into a disc around it, a disc containing only a little of the mass of the system but, thanks to its position far out from the centre, lots of the angular momentum.

From Dust to Pebbles

The planets began to form in the dusty cloud even before it had settled into a disc. The tiny grains of dust in the cloud collided with each other more and more frequently as the cloud collapsed, because the volume of the cloud ▷▷

3

3. As stars shrink they spin faster, like an ice-skater pulling in her arms.

FORMATION OF THE SOLAR SYSTEM

1. Collapsing cloud of gas gets hot inside

5. Gravity becomes important as the lumps get bigger

6. Gravity pulls the large lumps into spherical shapes

2. As the star forms, matter settles into a disc around it

3. Matter in the disc begins to clump together

4. Rocks form around the star

7. Planets like the Earth coalesce out of the rocky material

1. Family portrait representing the planets in our Solar System (excepting Pluto).

2. The Solar System also contains many smaller pieces of material – cosmic rubble such as comets.

was getting smaller, and there was less space between the grains. These very gentle collisions allowed the grains to stick together to make fluffy particles a few millimetres across, and these supergrains then collided with other supergrains to make bigger particles. Eventually, so many grains stuck together that objects the size of pebbles, then bigger lumps of rock formed, and gravity began to play its part. Once the primordial rocks were big enough, gravity would have pulled them together, and then the biggest rocks, by now kilometres across, would begin to dominate, attracting smaller lumps of rubble by gravity and growing bigger still.

As the lumps of rock grew, they collided with other lumps of rock. But because all the proto-planets, as they now were, were moving in the same direction around the Sun, these collisions would be relatively gentle, allowing the rocky lumps to join together, rather than be blasted back into dust. The larger lumps in the dusty disc grew to become planets, sweeping up most of the cosmic junk that still littered the disc. There is a reminder of those days still present in our Solar System. Between the orbits of the planets Mars and Jupiter, there is a band of rubble called the asteroid belt, populated by objects thought to be leftover material from the era of planet formation. There are estimated to be more than a million rocks in the asteroid belt bigger than 1 km across, with countless smaller lumps. The biggest, Ceres, is 933 km in diameter. But the total mass of all those rocks put together is only 15 per cent of the mass of our Moon.

Two Kinds of Planet

This picture of how planets formed neatly explains why there are two main kinds of planet in our Solar System. Close to the Sun, its heat would have driven away volatile material, leaving solid balls of rock with only thin atmospheres, such as Mercury, Venus, Earth and Mars. Further out from the Sun, it would have been cool enough for the rocky cores of planets to hold onto large amounts of gas, forming the giant planets Jupiter, Saturn, Uranus and Neptune. And further out still, there was scope for frozen balls of ice to remain in orbit around the Sun and that's where we find the ice planet Pluto and the comets that occasionally fall into our part of the Solar System from the outer fringes.

Recently, though, this simple picture has been challenged by the discovery of Jupiter-like

2

DUSTY DISCS

The astronomical explanation of how planetary systems form is no longer pure speculation, but is based on observations made in the 1990s of dusty discs of material around many young stars. The best studied of these discs is associated with a star called Beta Pictoris (right), and covers a span of at least 1000 AU. This is huge compared to the Solar System (the outermost giant planet, Neptune, is only 30 AU from the Sun), and shows that the disc is in the early stages of settling down, with some material still being ejected and blown away into space.

The Beta Pictoris system is thought to be only about 200 million years old, and the mass of material in the disc today is about one and a half times the mass of the Sun. Most of this will be lost as the system settles down. The inner part of the disc, comparable in size to our Solar System, is warped and distorted, possibly by the gravitational influence of planets forming within it.

The Hubble Space Telescope discovered hundreds of discs like this around young stars. This discovery is of key importance in our understanding of how planets form, and the fact that there are dusty discs around so many young stars suggests that planet formation is a common process. Further investigations of these objects will tell us what the Sun was like when the planets formed.

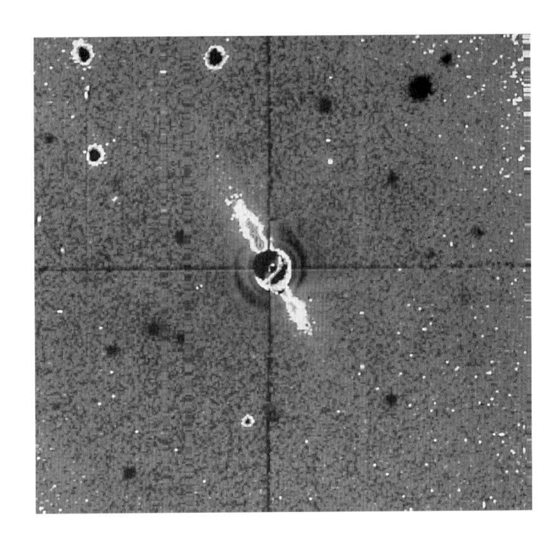

giant planets in Mercury-like orbits around other stars. Astronomers do not yet understand exactly how systems like this can form, although the discoveries are further evidence that solar systems are common in the Universe.

Whatever the details, the important point is that planet formation is a natural consequence of the way a dusty cloud of material collapses to form a single star. However this doesn't necessarily mean that all stars have planets. Multiple star systems have no problem getting rid of angular momentum, because it is stored in their orbital motion around each other, and it is unlikely that there are stable planetary orbits in these more complicated systems. Rather more than half of all stars seem to occur in systems with at least one other stellar

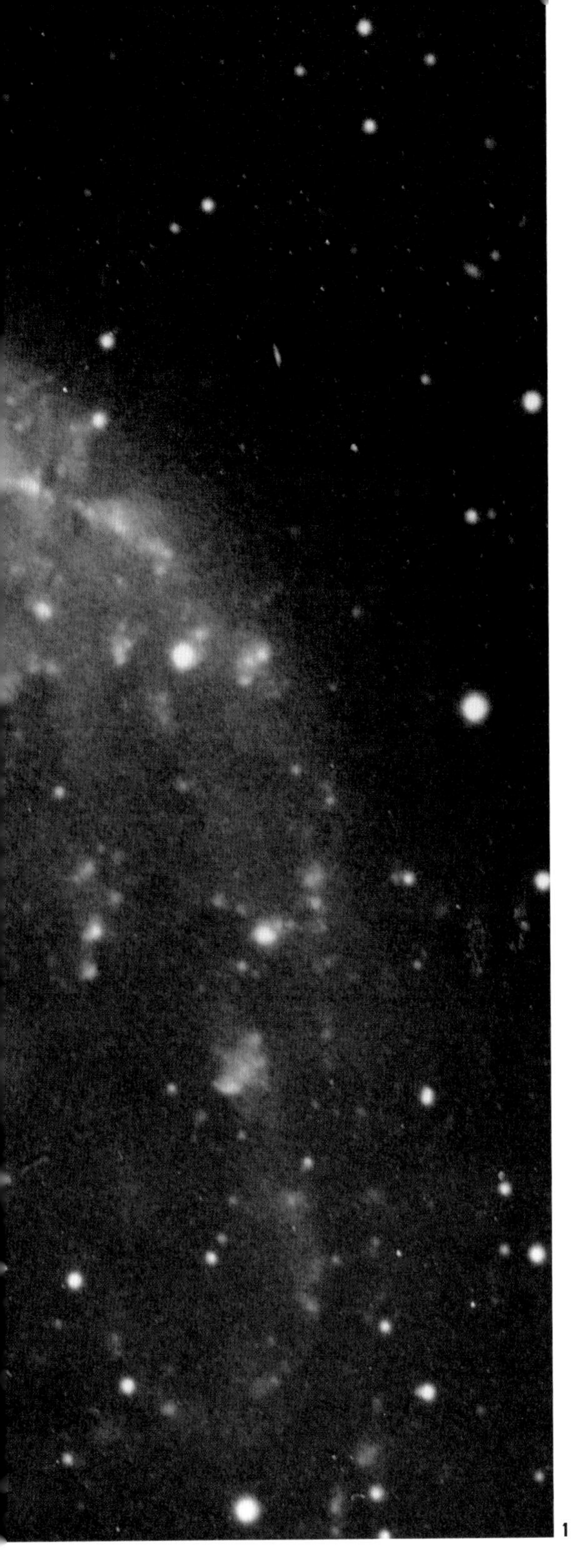

THE COSMIC SQUEEZE

Astronomers have a good understanding of how stars and planets form when a cloud of gas and dust in space begins to collapse. But what is it that triggers the collapse in the first place? In the Milky Way (and other disc galaxies) today, stars continue to be born because of the activity associated with the presence of other stars in the galactic disc itself.

The Orion Nebula is part of a much bigger cloud of gas, called a 'giant molecular cloud', which covers almost all of the constellation Orion on the night sky. Such giant molecular clouds are held together by gravity and can be regarded as single entities, the most massive single entities in our Galaxy, with a mass up to 10 million times the mass of the Sun and diameter of between 46 and 77 parsecs (150 and 250 light years). If a star explodes to create a supernova on one side of such a cloud, it sends out shock waves in the form of ripples moving through the cloud. These shock waves sweep up material in front of them (almost literally in the way that a broom sweeps dust into a heap), and this compresses some of the gas in the cloud enough to start it collapsing to form new stars with a whole range of masses. The biggest blobs of gas in these compressed regions collapse very quickly and form massive stars – tens of times as massive as our Sun which run through their life cycles very rapidly (in less than a million years) and explode in turn, sending more ripples out across the molecular cloud. In this way, a wave of star formation can travel right across a giant molecular cloud over a span of 10 or 20 million years.

companion (some have two or three). But even if just under half of the stars in the disc of the Milky Way are single systems like our Sun, that still leaves scope for well over 100 billion solar systems to have formed in our Galaxy in the way just described.

How Galaxies Make Stars

This is still not the complete picture, because the molecular clouds themselves are associated with the spiral arms of a galaxy like our own. They have come into existence because the thin

1. Planetary systems like our Solar System seem to be a natural part of spiral galaxies like the Milky Way.

gas between the stars has itself been squeezed on its journey around the disc of the Galaxy. Although the pattern made by the spiral arms twining outwards from the nucleus of a disc galaxy looks superficially rather like the pattern of cream stirred into a cup of coffee, there is one important difference. The spiral pattern made by the white cream against the black coffee quickly dissolves into a smooth brown colour as it mixes in. In the same way, the spiral pattern in a disc galaxy ought to get smeared out as the individual stars move on in their own orbits, at their own speeds around the centre. This should happen within about a billion years, a blink of an eye in the lifetime of a galaxy. But the spiral arms are not dissolved by rotation because they are constantly being renewed.

The distinctive pattern that shows up so clearly on photographs of spiral galaxies is caused by the presence of hot young stars along the edges of the spiral arms. These hot young stars are being born there because new clouds of gas and dust are constantly moving through the spiral pattern and being squeezed by a shock wave there. It is the shock wave that is the more or less permanent feature, like a sonic boom spiralling around the Galaxy, with the gas and dust just passing through. A useful way to picture what is going on is to think of a crowded motorway, where a slow-moving, large vehicle causes a kind of mobile traffic jam. Vehicles pile up behind the obstruction, work their way slowly past it, then speed up on the other side. The traffic jam moves along the road at a steady speed, but the individual cars that make up the jam are constantly changing as new cars approach from behind and others escape from the front of the jam. In the case of the spiral arms of our Galaxy, the shock wave itself is travelling around the Galaxy at a speed of about 30 km per second, while the stars and clouds of gas and dust are overtaking the spiral wave at a speed of about 250 km per second, passing through it but getting squeezed in the process. Clouds of gas and dust pile up in the cosmic traffic jam along the inside curve of the spiral arm, and are squeezed there, triggering bursts of star formation like the one going on in the Orion molecular cloud.

How Stars Make Stars

The whole process is self-sustaining. The shock waves are maintained by the stellar explosions going on along the edge of the spiral arm, and the stellar explosions are caused by the squeeze given to clouds of gas and dust by the shock waves. Although giant molecular clouds are constantly forming stars in this process, the explosions of other stars constantly feed raw material back into interstellar space to make new molecular clouds, the raw material for later generations of stars. Each year, just a few solar masses of material are being recycled in this way in the Milky Way Galaxy, but over a few billion years that adds up to a lot of new stars.

At any one time the amount of matter orbiting around the disc of the Galaxy in the form of giant molecular clouds is about 3 billion times as much as there is in our Sun. This is equivalent to about 15 per cent of the total mass of stars in the disc itself. It takes surprisingly few big stellar explosions to keep the recycling process going. There are only two or three supernova explosions in our Galaxy every century, and there hasn't been one close enough to be studied by human observers for nearly 400 years (the closest occurred in 1987 in a small nearby galaxy, the Large Magellanic Cloud). Think of this in terms of the lifetime of a galaxy. Even at a rate of only two per century, there will be 20,000 supernova explosions every million years, and several hundred million have occurred since the Sun was born 4.5 billion years ago.

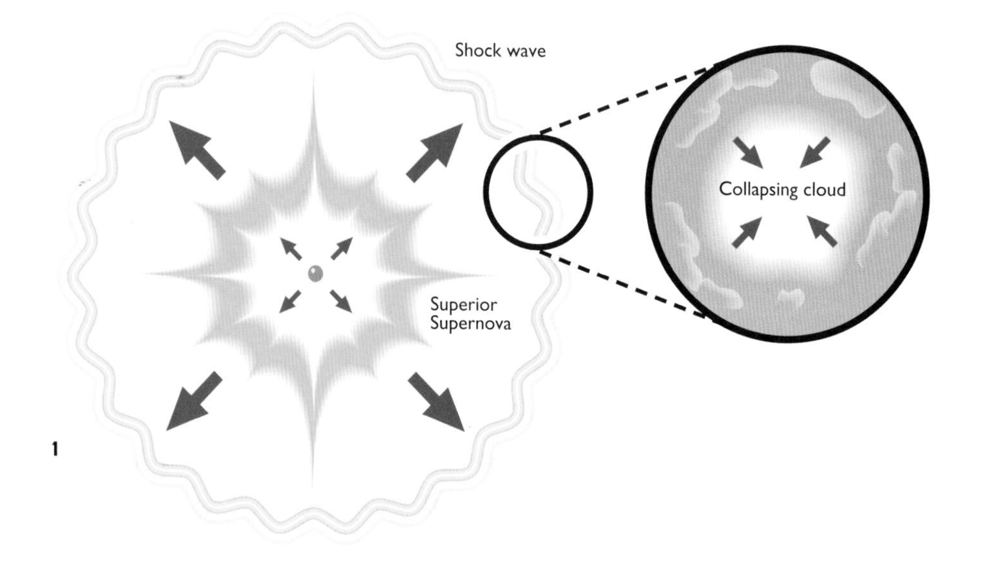

1. The expanding shockwave from an exploding star (main picture) squeezes clouds of gas and encourages new stars to form (inset).

2. (opposite) The large Magellanic Cloud is a near-neighbour of our galaxy. It is about 160,000 light years away, and 30,000 light years across.

COOKING THE ELEMENTS

All of the elements in the Universe today, except for hydrogen, helium and a very small amount of light elements such as lithium and deuterium, have been manufactured inside stars and spread through space late in the lifetimes of those stars, either as clouds of gas puffed away gently from the surface of the star or in great stellar explosions. The raw material for the manufacture of those elements is the hydrogen and helium which were produced in the Big Bang when the Universe was born. Some of that hydrogen is being converted into helium today inside main sequence stars, and more was converted into helium in the past. But by far the bulk of the 25 per cent helium that contributes to the composition of a star like the Sun is primordial stuff left over from the Big Bang. All the heavier elements, such as oxygen, carbon, nitrogen, iron, and all the elements in your body, are literally star dust created from this primordial matter.

The fundamental component of an element is, of course, the atom. All atoms of a particular element (such as oxygen) are identical to one another and are made up of a core (the nucleus) containing particles called protons and others called neutrons, surrounded by a cloud of electrons (one electron in the cloud for every proton in the nucleus). It is the number of protons in the nucleus which determines the nature of the

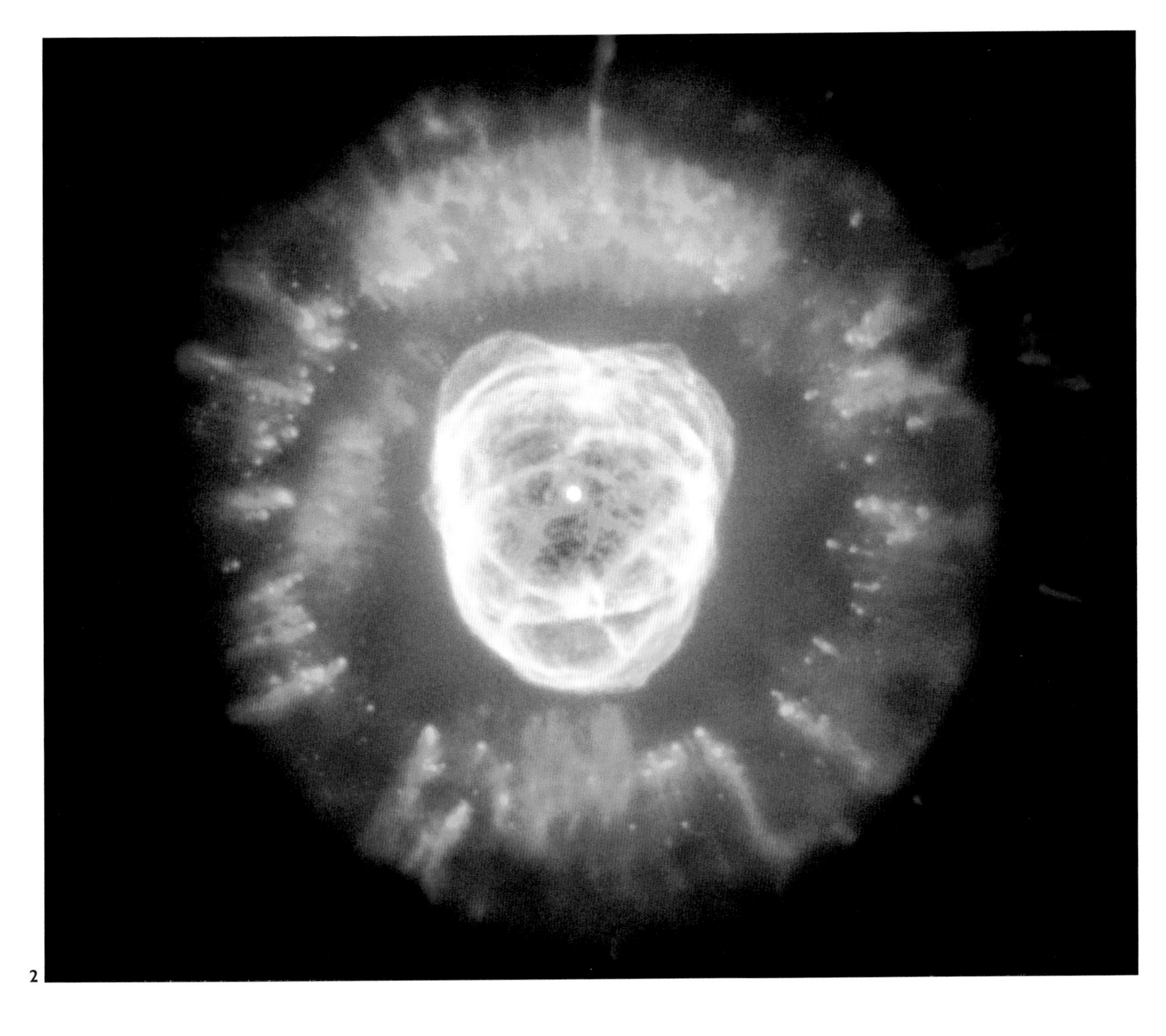

2

1. Eta Carinae is a star about 100 times as massive as our sun, blowing material away into space.

2. The Eskimo nebula is one of the prettiest examples of a planetary nebula.

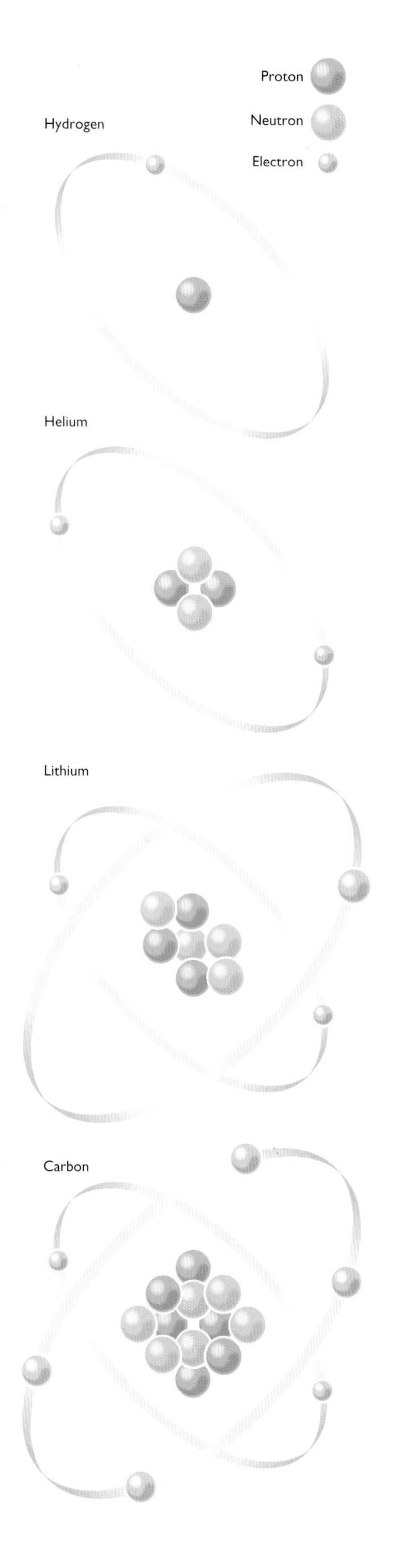

1

atom, whether it is an atom of lead, or sulphur, or whatever. The simplest atom of all, hydrogen, has a nucleus consisting of a single proton.

The Stellar Pressure Cooker

Under the extreme conditions that exist in the hearts of stars (temperatures in excess of 15 million K and densities greater than 10 times the density of lead on Earth), the electrons are stripped away from their atoms, and the nuclei are free to move about through a sea of free electrons and collide with one another. Very occasionally, the nuclei interact to make new nuclei, making new elements in the process known as stellar nucleosynthesis.

Making helium out of hydrogen is the first step in nucleosynthesis. Although this mostly happened in the Big Bang, the stellar process is still important in providing the energy that keeps main sequence stars shining. The story of how elements are cooked inside stars really begins later in a star's life, when conditions become extreme enough for helium nuclei to combine with one another to make nuclei of carbon.

The problem with nucleosynthesis is that it is difficult to make nuclei stick together. All nuclei carry positive electric charge, because protons are positively charged and neutrons have no charge at all. If the nuclei can come close enough together (in effect if they touch) then they can interact and (sometimes) fuse to make new nuclei, held together by a strong nuclear force which overcomes the electrical repulsion but which has only a very short range. Slow-moving nuclei are repelled from one another electrically before they can interact in this way; only fast nuclei (which means hot nuclei) can collide powerfully enough to overcome the electrical repulsion. The more protons there are in the nucleus, the stronger the repulsion is and the hotter the nuclei must be before they can interact.

The extra heat required to initiate stellar nucleosynthesis is provided, paradoxical though it may seem, when the core of a main sequence star has used up all of its hydrogen fuel. Because the core can no longer generate heat by converting hydrogen into helium, the temperature and pressure there fall, and the weight of the star pressing down on the core makes it collapse. But as the core begins to collapse, gravitational energy is released, making the core hotter again. It gets so hot that helium nuclei can move fast enough to stick together. Energy is released (through $E = mc^2$ ▷ p. 58) as they do so and the collapse is halted for as long as the supply of helium lasts.

Cooking up Carbon

Each helium nucleus involved in this process contains two protons and two neutrons (and is known, for obvious reasons, as helium-4). (There is another variety of helium containing two protons and a single neutron (helium-3), but it is not important here.) You might think that the natural result of an interaction between two helium-4 nuclei would be a nucleus containing four protons and four neutrons (beryllium-8). But it turns out that this nucleus is very unstable and breaks apart within 10 millionths of a billionth of a second after it forms. The only way that helium-4 nuclei can combine to form a new stable nucleus is if three of them get together at very nearly the same time to make a nucleus of carbon-12. This step is made much easier by the fact that the carbon-12 nucleus has just the right energy to be stable. The three helium nuclei initially form what is called an 'excited' carbon-12 nucleus, which then gets rid of the excess energy by radiating it away, forming a stable nucleus. This is called the 'triple alpha process'.

Just as a helium-4 nucleus has slightly less

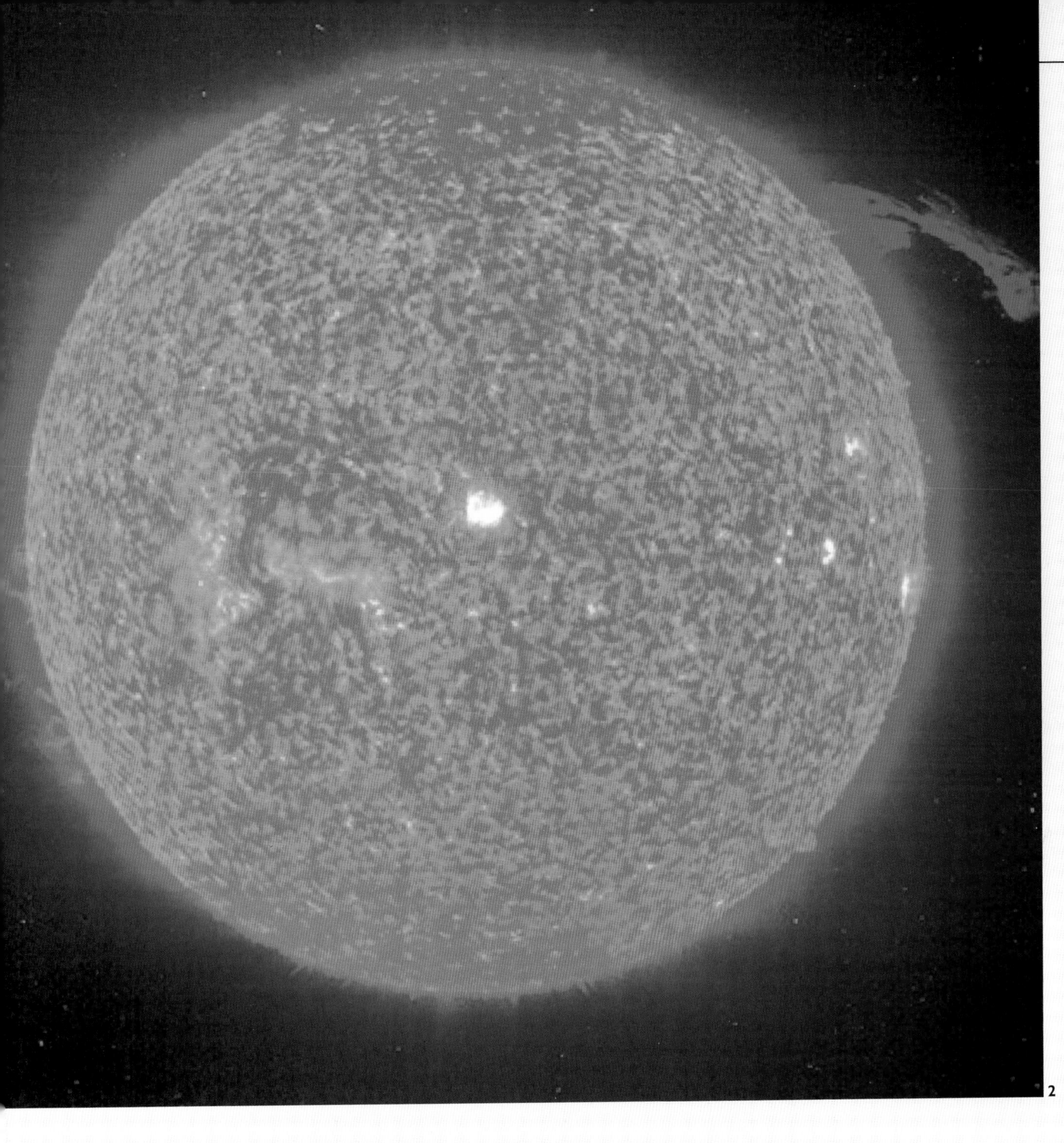

1. The p–p chain, which releases heat inside the Sun.

2. A SOHO ultraviolet image of the chromosphere layer of the Sun's atmosphere, with a solar flare pictured top right.

(converting the nucleus itself into carbon-13). If a second proton penetrates this carbon-13 nucleus, it becomes nitrogen-14. A third proton will convert it into oxygen-15, which is unstable and spits out a positron and a neutron as it converts into nitrogen-15.

Now comes the crunch. If a proton penetrates a nucleus of nitrogen-15, it spits out a whole alpha particle (a helium-4 nucleus), leaving behind a nucleus of carbon-12 identical to the one that started the cycle. The net effect is that four protons have been turned into one helium-4 nucleus with energy being released along the way. As with the p–p chain, 0.7 per cent of the mass of the set of four protons is converted into energy each time an alpha particle is made.

TOPIC LINKS

mainly of hydrogen and helium. Red giants can have masses tens of times the mass of the Sun, and diameters up to 100 times larger. Because they are so big, the gravitational pull at the surface of such a star is very weak and it is easy for material to escape into space. This is especially true when the atmosphere is being puffed in and out by variations like those we see in Cepheids.

The Fate of the Sun

The Sun itself will become a red giant in about 5 billion years' time, and at its largest it will expand to engulf Mercury and approach the orbit of Venus. You may sometimes hear or read that it will even engulf the Earth, but this prediction is not correct, because it does not take account of the fact that by that stage of its life the Sun will have lost about 25 per cent of its original mass by ejecting material into space.

The overall time that a star spends as a red giant is much less than the time it spends on the main sequence, only between 5 per cent and 20 per cent of its main sequence lifetime, depending on its mass. The Sun will be a red giant for only about a billion years, and will never get beyond the helium-burning stage. Bigger stars, though, may go through successive phases of nuclear burning, ending up with a structure like that of an onion, with different kinds of nuclear burning (and nucleosynthesis) going on in each layer.

Fading Stars

For a star like the Sun, and stars with up to a few times as much mass as the Sun, all the nuclear burning possibilities will eventually come to an end. The star will blow away its outer layers to form a planetary nebula as the core collapses and stabilizes as a solid lump of material. This dense core of star stuff starts out hot, thanks to the leftover heat of its former glory and the heat generated in its final collapse, but it is very small, about the size of the Earth. It becomes what is known as a 'white dwarf', one of the hot but faint stars that occupy the bottom left part of the HR diagram.

A white dwarf contains anything from about half to one and a half times the mass of our Sun, packed into a solid lump about the size of the Earth. One cubic centimetre of white dwarf stuff would have a mass of about 1 tonne – a million times the density of water.

If all stars faded away quietly, as the Sun is destined to do, there would be very little in the way of heavy elements in the Universe, very few planets (if any) and no life forms like ourselves. But some stars end their lives in events that play a key role in both manufacturing heavy elements and recycling them into the interstellar medium.

However, stars which start their lives with more than about eight times as much mass as our Sun have a different, and more spectacular, fate in store. They are destined to become the supernovae which trigger the next generation of star birth, allowing new stars to arise Phoenix-like from the ashes of the old.

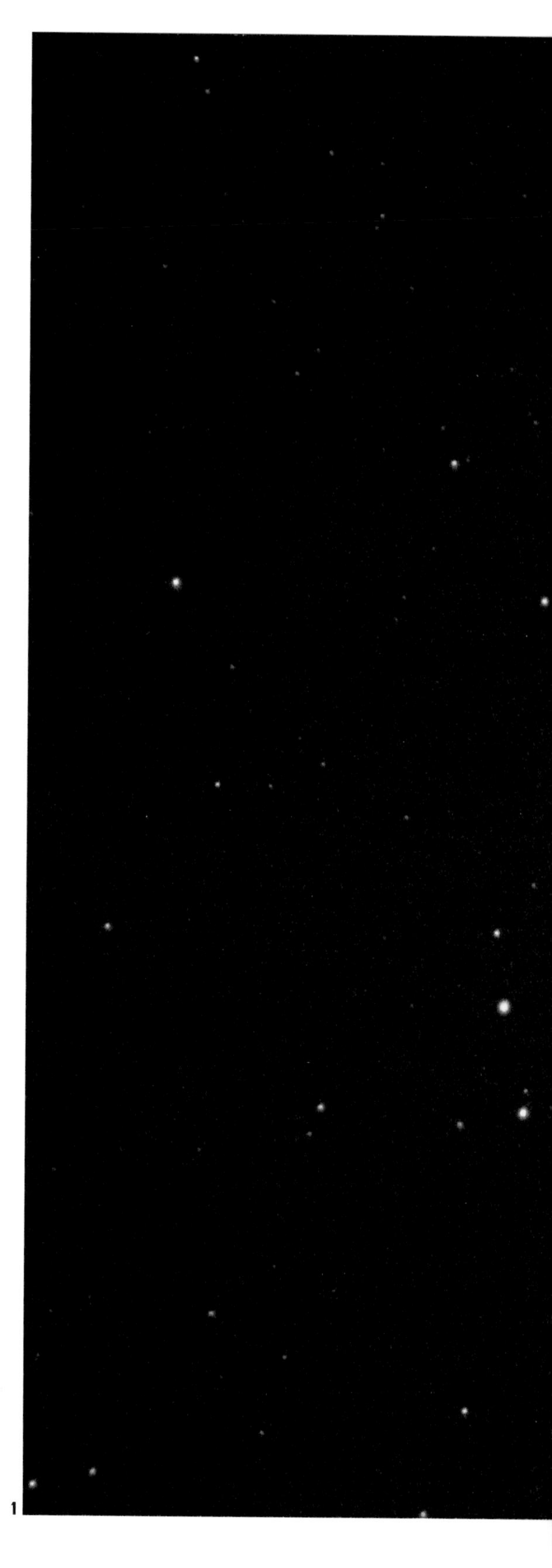

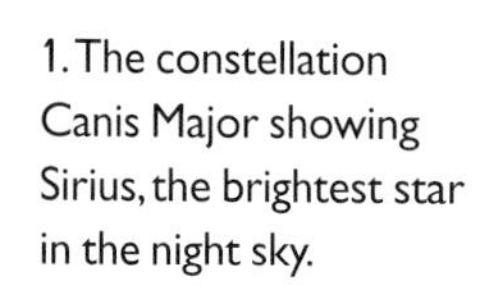

1. The constellation Canis Major showing Sirius, the brightest star in the night sky.

A BIGGER BLAST

Type II supernovae occur in massive, young stars rich in heavy elements produced by nucleosynthesis. These stellar explosions mainly occur in the spiral arms of disc galaxies (▷ p. 22), because the stars involved are so massive they do not have time to move far from their place of birth before they die. They also occur in other regions where starbirth has been triggered. This happens, for example, when clouds of gas and dust in a relatively quiet galaxy are disturbed by the tidal forces created when another galaxy passes nearby, and collapse to form new stars. Type II supernovae release even more energy when they explode than Type I supernovae, but they are not as bright to conventional telescopes, because most of this energy is released in the form of invisible particles called neutrinos. A Type II supernova produces in a few minutes about 100 times as much energy as the Sun will radiate over its entire lifetime of 10 billion years or more.

The more massive a star is, the more quickly it burns its fuel, and the shorter its life. The progenitors of Type II supernovae can have masses tens of times that of our Sun, but to give an example of how they are formed we must look at the evolution of a star which starts out with a little less than 20 solar masses of material. It is thought to be the mass of a supernova that was seen to explode in the Large Magellanic Cloud in 1987, the supernova known as SN 1987A.

Running Out of Fuel

Such a star has to burn nuclear fuel so fiercely to hold itself up that it shines 40,000 times brighter than our Sun and spends only 10 million years on the main sequence. Burning helium as a red giant supplies it with energy for only another million years or so, and then it runs through the remaining possibilities offered by nuclear fusion at an ever faster rate. Converting carbon into a mixture of oxygen, neon and magnesium provides energy for 12,000 years; burning neon and oxygen sustains it for another 16 years; and fusion of silicon nuclei to make iron-family nuclei keeps it going for about a week. During that last week of its life as a more or less stable star, the inner core of the giant star is like a series of Russian dolls, with each of these nuclear

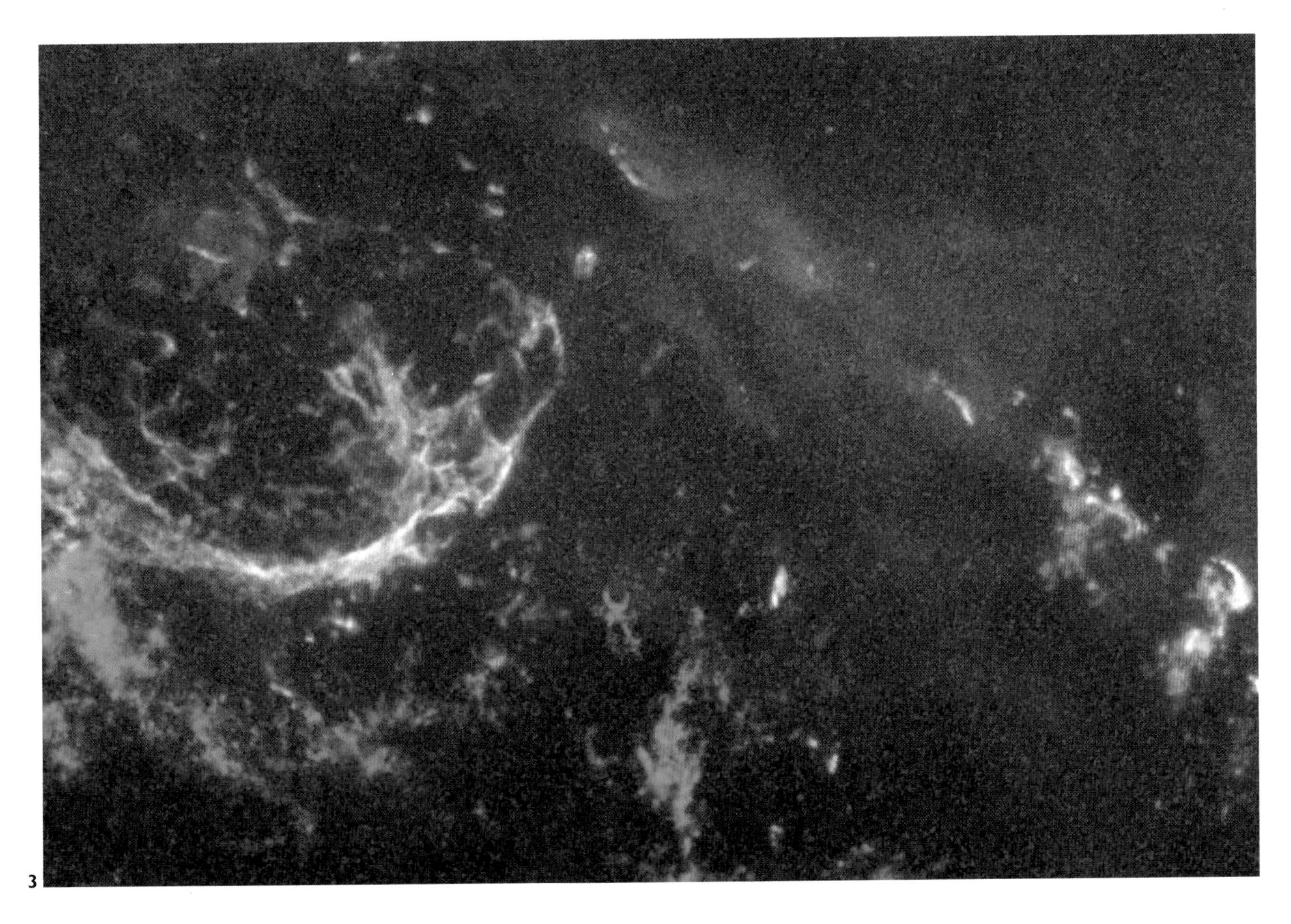

3. The remnant of a supernova that exploded long ago in the Large Magellanic Cloud. The colour indicates the presence of large amounts of oxygen.

1. One of our nearby irregular galaxies, NGC 1313.

2. Relative sizes of stars. To help put this into perspective, a white dwarf is about the same size as Earth.

fusion processes going on one inside the other.

Once the silicon in the core has been converted into iron-family elements, there is no longer a source of energy to provide pressure to hold the star up against its own weight. With the rug pulled from under it, the star collapses in spectacular fashion, converting gravitational energy into a heat so intense that it breaks the heavy nuclei apart. This produces a pressure so great that it forces electrons to combine with protons to make neutrons. The inner core of the star collapses in a few seconds from a sphere of star-stuff bigger than the Sun into a ball of neutrons about 20 km across. This leaves the outer layers of the star, with perhaps 15 times as much mass as our Sun (remember that it will have lost a lot of its original mass during its time as a red giant), plunging inward at about a quarter of the speed of light. But the formation of the neutron star produces a shock wave which ripples out from the core in a kind of rebound, rapidly followed (really

4

1. Actual photograph of the centre of a system like the ones illustrated here. This is a black hole at the heart of the galaxy M51 taken by the Hubble Space Telescope.

2. Illustration of a stellar-mass black hole ejecting jets of material from its poles.

3 and 4. Close-up views (artist's impression) of black hole activity.

effect – ▷ p.120) tells us how big the central black holes in these galaxies are.

In a typical example, the galaxy NGC 7052, there is a disc of material 1100 parsecs (3700 light years) in diameter swirling round a black hole with a mass 300 million times the mass of our Sun. The disc itself contains 3 million solar masses of material – enough to keep the quasar shining for 3 million years, because it swallows only one solar mass a year.

Bending Space

The general theory of relativity explains gravity as a result of space being bent by the presence of matter. In a neat aphorism, physicists say that matter tells space how to bend; space tells matter how to move. In effect, objects moving under the influence of gravity roll along the valleys in a hilly landscape of curved space (strictly, curved spacetime). Using this image, flat spacetime, without any matter in it, is like a stretched flat sheet of rubber (or a trampoline). A heavy weight placed on the stretched sheet makes a dent, and anything rolling across the sheet follows a curved path around the dent. If the object is heavy enough, it stretches the fabric so much that it makes a deep pit with vertical sides, from which nothing can escape – a black hole.

TOPIC LINKS

1.4 Out with a Bang
p.71 Black Holes and Neutron Stars
p.75 Proof of the Pudding

2.1 The Big Bang
p.87 The Singular Beginning

4.3 Into the Unknown
p.217 Beyond the Black Hole
p.230 The Evolution of Universes

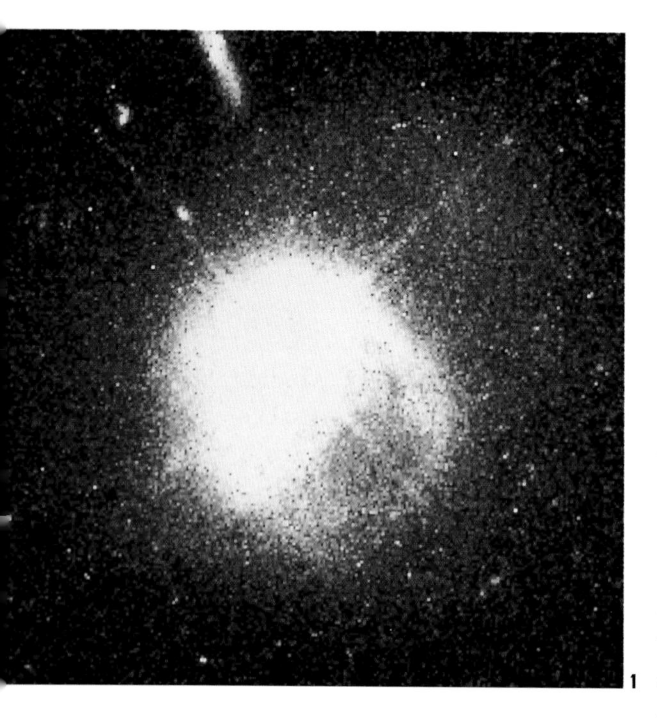
1

Gamma ray bursts from space were first identified in the late 1960s by American satellites designed to look for gamma rays from any secret nuclear bomb tests being carried out in Russia.

Singularities

The discovery of pulsars meant that the equations which said neutron stars exist had to be taken seriously. That meant that another, even stranger, prediction of those equations also had to be taken seriously. At the end of the 1930s, Robert Oppenheimer and George Volkoff had shown that the equations which say neutron stars exist also say that there is an upper limit to the amount of mass a neutron star can have. The exact value of this upper limit depends on subtleties in the equations, which the experts still argue about, but it is about three times the mass of the Sun. What would happen to any neutron star that tried to form with more mass than this, or which started out with less mass but crossed the line as a result of accreting material from a companion? The equations said that it would shrink down indefinitely, towards a single point – a singularity.

But on the way to a singularity, a collapsing object would disappear from view, because the gravitational pull at its surface would become so intense that nothing, not even light, could escape. The work of Robert Oppenheimer and his colleagues was not followed up at the time because of the Second World War. Oppenheimer himself worked on the Manhattan Project – to build the first nuclear bomb.

In fact, collapsed objects like this had already been described, using the equations of Albert Einstein's general theory of relativity, by Karl Schwarzschild, in the second decade of the twentieth century. In terms of relativity theory, the collapsed object is pinched off from our Universe because space itself (strictly speaking, 'spacetime') has been bent (or warped) around it, making a hole in space. The interior of the hole is, in effect, a separate, self-contained universe. These objects were only given their now-familiar name, 'black holes', in 1967, just after the discovery of pulsars made everyone begin to take the idea seriously.

Getting to Grips with Black Holes

Any object will form a black hole if it is squeezed into a small enough volume. For a particular mass, the crucial radius at which this occurs is called the Schwarzschild radius, and this becomes, in effect, the radius of the black hole. Once anything is squeezed within its Schwarzschild radius, it will collapse to a singularity, leaving a black hole with that radius as its imprint on our Universe, like the grin on the face of the fading Cheshire cat. For the Sun, the Schwarzschild radius is just 2.9 km, which shows how close neutron stars (with radii of about 10 km) are to becoming black holes. For the Earth, it is only 0.88 cm. But black holes need not necessarily be associated with very high densities. An object with the same density as the Sun (or water) would be a black hole if it were about as big across as our Solar System.

There is now direct proof that black holes exist. Even though nothing can escape from a black hole (so it cannot be seen directly), there may be a great deal of activity going on just outside the Schwarzschild radius (which is also known as the 'event horizon'). If a black hole with a radius of a few kilometres is in orbit around another star, stripping matter away from it and swallowing it up, there will be an intensely swirling maelstrom of stuff orbiting around the black hole in a disc and being sucked in. Gravitational energy released as the matter falls into the black hole will heat this disc to the point where it radiates X-rays. Such X-ray stars have been detected orbiting ordinary stars, and their masses, sometimes exceeding 10 solar masses, have been measured from studies of their orbits.

There is no doubt that some of these X-ray stars are black holes.

Much more massive black holes (perhaps containing 100 million solar masses of material, and as big across as our Solar System) are also thought to reside at the hearts of galaxies, where the energy released as they swallow matter makes the gas around some of them shine more brightly than the rest of the galaxy put together. They are known as quasars. To give you some idea of the enormous amount of energy that can be released when matter falls into a black hole, the energy output of a quasar can be maintained by swallowing only about one solar mass of material each year.

Proof of the Pudding

The best evidence that black holes exist came in the 1990s, when astronomers identified the sources of intense bursts of gamma rays detected by instruments on board satellites orbiting the Earth. These gamma ray bursts had long been known, but it was only at the end of 1997 that astronomers were first able to identify the source of one of these bursts using ordinary telescopes. It turned out to be in a galaxy more than 3 billion parsecs (10 billion light years) from Earth. In order to produce a burst of gamma rays visible to our detectors that far away, for a few seconds the object had radiated as much energy as every star in every galaxy in the visible Universe put together. In a region about 150 km across, it briefly produced conditions like those which existed in the Big Bang itself. The only way in which to generate such enormous outbursts of energy in such a short time is in a kind of super supernova (a hypernova?) where the core collapse does not stop at the neutron star stage, but goes all the way to a black hole. The power of a quasar (as bright as several hundred billion Suns) comes from swallowing about one solar mass of material in a year; the power of a gamma ray burster comes from swallowing several solar masses of material in a few seconds. It is the ultimate form of stellar death.

2

1. The quasar PKS2349 photographed by the Hubble Space Telescope.

2. Radiation coming from a region of space where a black hole has collapsed is like the fading grin of the Cheshire cat.

2

THE FATE OF THE UNIVERSE

2.1 THE BIG BANG

One of the greatest achievements of the human intellect was the discovery in the twentieth century that the Universe as we know it originated in a hot, superdense state at a definite moment in time (a beginning, if you like), that it has been expanding ever since, and that we can work out when time began – about 14 billion years ago. It is pleasing too that the age of the Universe, determined from cosmology, closely matches the ages of the oldest stars determined by astrophysics.

There is overwhelming evidence in support of both the idea of the Big Bang, as it is known, and when it occurred. At the beginning of the twenty-first century, instead of resting on their laurels, cosmologists are now attempting to answer the other big questions about the Universe: where it is going and how it will all end. The first tantalizing clues that we might soon be able to give definite answers to these questions came at the end of the twentieth century.

Previous page. The Hubble Deep Field. Galaxies seen by light that left them more than 10 billion years ago.

THE LAW ACCORDING TO HUBBLE

The most important thing we know about the Universe is that it is expanding, that galaxies (or rather whole clusters of galaxies) are getting farther apart as time passes. We cannot see the distance increasing between them, because the distance scales and time scales involved are so huge. Even if we watched for a million years, we would scarcely be able to detect the expansion of the Universe directly. But we know for sure that the Universe is expanding, because we can measure both the distances to many galaxies and the speeds with which those same galaxies seem to be receding from us.

The key discovery is that there is a very simple relationship between these two properties, telling us that the apparent velocity recession is proportional to the distance of a galaxy from us. This is known as Hubble's Law, and the constant of proportionality in Hubble's Law is called the Hubble Parameter (or sometimes, Hubble's Constant); it is denoted by the letter H. Hubble's Law does not mean, however, that we are at the centre of the Universe. It is the only velocity/distance law (except for a situation in which nothing is moving at all) which holds wherever you are in the Universe.

Everything is receding from everything else, like raisins being carried away from one another in the rising dough of a raisin loaf as it

To a cosmologist, a galaxy like the Milky Way, containing hundreds of billions of stars, is the *smallest* thing in the Universe worth taking account of.

1. The Andromeda galaxy, also known as M31, is the nearest large spiral galaxy to our own.

cooks. Whichever galaxy you sit on, it will look as if all the other galaxies are rushing away from you with speeds proportional to their distances.

Beyond the Local Group

The first measurements of distances to galaxies beyond the Milky Way were made at the end of the 1920s, using the Cepheid distance scale (▷pp. 28–9). But even with the aid of the best telescopes that were available before the 1990s, it was not possible to detect Cepheids in more than a handful of the very nearest galaxies to our own. However, this was enough to show that the Milky Way and the Andromeda galaxy (also known as M31) are the two largest members of a small group of galaxies which also includes the Large and Small Magellanic Clouds, and is collectively known as the Local Group. The Local Group is actually a very small cluster of galaxies – other clusters of galaxies contain hundreds or even thousands of individual galaxies. In a cluster, the individual members are held together by gravity, and move around within the cluster like individual bees moving within a swarm. But the cluster as a whole takes part in the expansion of the Universe (the swarm of bees moves as a unit).

Right up until the 1990s, measuring distances to galaxies beyond the Local Group depended on using so-called 'secondary indicators' calibrated within the Local Group. Because the distances to galaxies in the Local Group (in particular, the distance to the Andromeda galaxy) are known from the Cepheid method astronomers can study the apparent brightnesses of bright objects in these nearby galaxies and use the known distances to work out how bright they really are. This makes it possible to calibrate the brightness of things like globular clusters, supernovae, and the huge star-forming clouds known as 'HII regions'. Then, by identifying

THE OLDEST THINGS IN THE UNIVERSE

Until the middle of the 1990s, astronomers were slightly embarrassed to admit that their best estimates of the age of the Universe came out at slightly less than their best estimates of the ages of the oldest stars. Obviously, the Universe must be older than the stars it contains, but measurements of the Hubble Parameter using ground-based telescopes gave rather rough and ready estimates. They showed that the Universe was about 10–12 billion years old, while the oldest stars were thought to be 14–15 billion years old. The astronomers weren't too perturbed, however, because both measurements were difficult and plagued with uncertainties, and they expected one or both of them to need adjustment when telescopes were flown above the obscuring layer of the Earth's atmosphere.

In fact, both of them needed adjusting, and in the 'right' way. In the second half of the 1990s, data from the Hubble Space Telescope (left) showed that the Hubble Parameter is a little smaller than had been thought. This meant that the best estimate for the age of the Universe became 14 billion years. At almost the same time, data from the HIPPARCOS satellite showed that some of the stars used to calibrate stellar ages were a little further away than had been thought, so they must be a little brighter than was thought in order to look as bright as they do to us. And that implies they must burn their fuel faster than anyone had thought, so they haven't taken as long to get to their present state. The best estimate for the ages of the oldest stars came down, as a result, to 12–13 billion years.

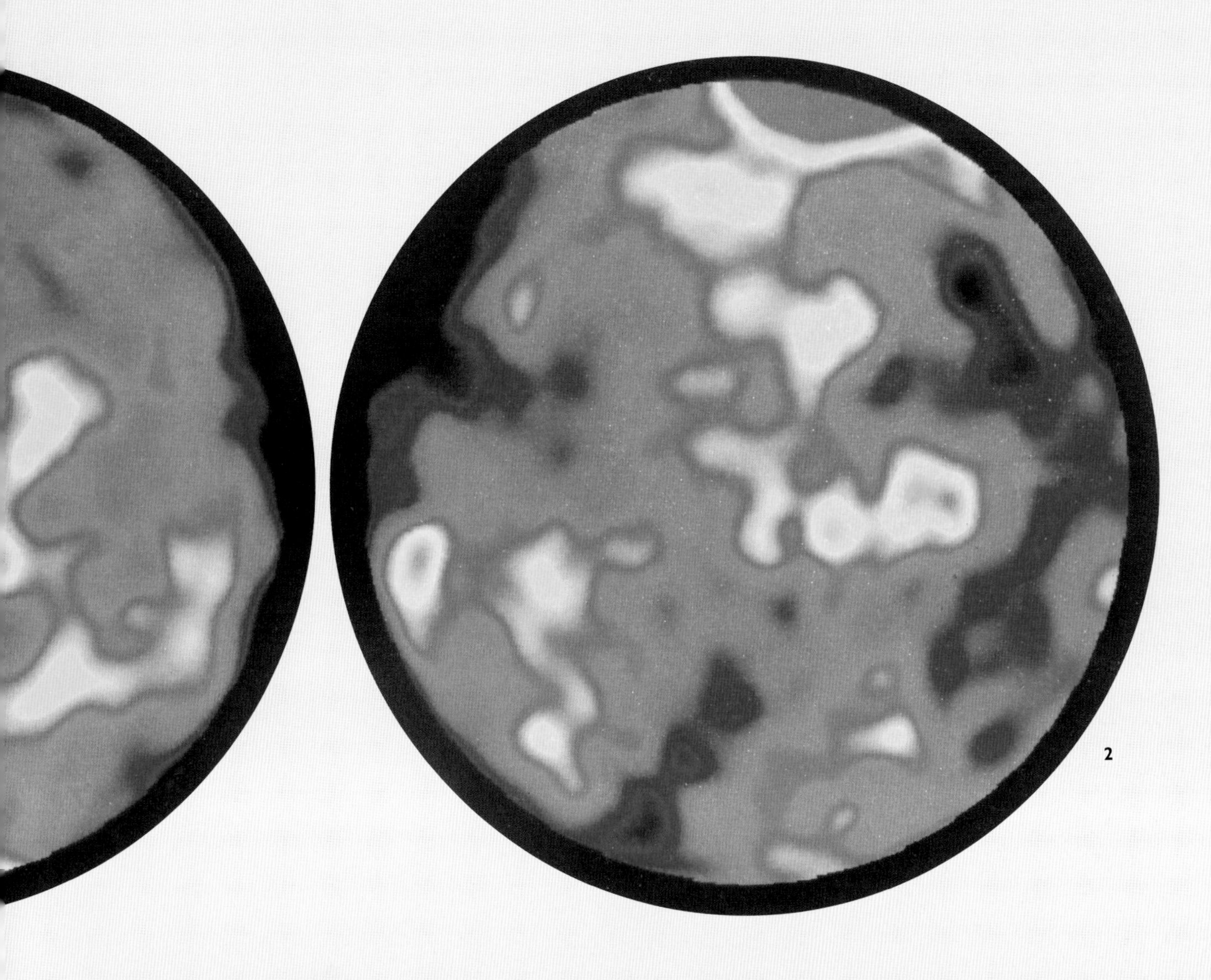

2. Detailed map of temperature variations when the Universe was half a million years old, based on data gathered by COBE over four years.

zero, the temperature had fallen to a billion K, just six times hotter than the heart of the Sun today, and the proportion of neutrons to protons had fallen to just 14 per cent. But they were saved from disappearing entirely because at last it was cool enough for nuclei of deuterium, and other light nuclei, to stick together permanently.

Saving Neutrons

In a flurry of nuclear reactions over the next few seconds, almost all the remaining neutrons in the Universe were locked up with protons in nuclei of helium-4, producing a mixture of about 25 per cent helium (in terms of mass), nearly 75 per cent hydrogen, and a trace of very light elements such as deuterium and lithium. This process of Big Bang nucleosynthesis finished about 3 minutes and 46 seconds after time zero – in round terms, four minutes after the beginning.

The standard Big Bang theory predicts exactly the proportion of these light elements actually seen in the oldest stars.

TOPIC LINKS

1.3 Stellar Evolution
p. 53 Cooking the Elements

2.2 Cosmology for Beginners
p. 94 The Steady State Mode

2.4 The Accelerating Universe
p. 132 Quantum Fluctuation
pp. 134–5 Balloons and the Background Radiation

2.2 COSMOLOGY FOR BEGINNERS

Cosmology started out as a game played by mathematicians who liked to tinker with the equations of the general theory of relativity, which describes the behaviour of space, time and matter. The Universe is the biggest collection of space, time and matter they could think of. The game started soon after Albert Einstein discovered the general theory, in 1916. Before then, any views that people had about the origin of the Universe and its ultimate fate depended entirely on philosophy or religion, and had no scientific basis.

After Einstein's discovery, ideas about the evolution of the Universe abounded, based on his equations. But it was impossible to say which set of ideas matched the real Universe. It was only right at the end of the twentieth century that cosmologists were able to test their calculations with accurate measurements of the way the real Universe behaves.

A VARIETY OF UNIVERSES

Mathematicians can use the equations of the general theory to provide mathematical descriptions of different ways in which space and time interact in the presence of matter – that is, different ways in which a universe could change as time passes. Notice the use of the lower case 'u' on 'universe' there. Cosmologists reserve the capital, Universe, to refer to the real version of spacetime in which we live. They refer to their hypothetical mathematical versions as universes, or as 'models'. There are very many (perhaps an infinite number) of model universes that can be described using Einstein's equations, and the big trick is to find one that looks just like our own Universe.

Fortunately, our Universe turns out to be a very simple variation on the theme, and can be described using a very simple version of the equations. Einstein himself marvelled at this, saying, 'the most incomprehensible thing about the Universe is that it is comprehensible.'

But just how simple *is* the Universe?

Simple Models of Space and Time

The most basic division of cosmological models into categories can be described in terms of the expansion of the Universe. The Universe is expanding today but the gravity of all the stuff in the Universe is tugging on all the stuff in the Universe and trying to slow the expansion down.

The difference between the two main kinds of model can be pictured by imagining the difference between a baseball being hit and a rocket being launched. Even the most powerful baseball player in the world cannot hit the ball hard enough for it to escape from the pull of the Earth's gravity. No matter how high the ball

1

2

1 and 2. No baseball player can hit the ball so hard that it never falls back to Earth. You need a very powerful rocket to escape the Earth's gravitational pull.

flies, it will slow down, stop, and then fall back to Earth. But a sufficiently powerful rocket can launch at such a high speed that it can escape from the bonds imposed by the Earth's gravity, leave the vicinity of the Earth entirely, and never return. It is said to have achieved 'escape velocity' from Earth – a velocity which depends only on the mass of the Earth itself.

The question is, is the Universe expanding fast enough to escape from its own gravity – will it expand forever, or one day slow down and stop like the baseball and re-collapse? It is relatively easy to measure how fast it is expanding, but much harder to measure how much mass there is in the Universe, so it took a long time to find the answer.

Two Complications

This simple picture is not really quite so simple.

First, Einstein's equations also allow for the presence of a number, called the 'cosmological constant', which affects the expansion rate. It is given the Greek letter lambda (Λ) in those equations, but nothing in the equations tells us what value lambda has. Depending on its size, it could act as a kind of antigravity, making the Universe expand faster, or as an extra gravitational influence, slowing down the expansion. This is great fun for the mathematicians, because it gives them lots more models to play with. But studies of the expansion of the real Universe show that even if the lambda term does exist it must be very small, and until the 1990s astronomers usually set the constant as zero, to make life easier.

The other complication is more of an oddity than a complication. If you could throw a ball upwards at *exactly* the escape velocity, and nothing got in the way, it would keep going forever, but it would keep slowing down forever, and after a very long time it would seem to be hovering, high above the Earth (strictly speaking, infinitely far away), never to fall back. This curiosity is interesting, because the Universe itself seems to be very nearly in this state. It can be pictured another way, using geometry.

THE STEADY STATE MODEL

In the 1940s, three astronomers (Fred Hoyle, Tommy Gold and, right, Herman Bondi), came up with an idea to explain the expansion of the Universe without invoking a Big Bang. They pointed out that if new atoms of hydrogen were being created by expanding spacetime at a rate of just one new atom in every 10 billion cubic metres of space each year, enough atoms would be produced to make new galaxies to fill in the gaps as the old galaxies moved apart. At any moment in time (any cosmic epoch), the overall appearance would be much the same as it is in the Universe today. This became known as the 'Steady State' model.

This is a good example of how astronomers explore the Universe in their imagination, by thinking up models that might correspond to reality. When people argued that the continual creation of matter seemed rather an extravagant hypothesis, the supporters of the Steady State model pointed out that it didn't seem any more extravagant than creating all the matter all at once in a Big Bang.

The rivalry between the Big Bang and Steady State models encouraged astronomers to probe the Universe with radio and optical telescopes to find out which one was right. Eventually, this probing produced clear evidence that the Universe has changed as time has passed. It is not in a steady state. The discovery and investigation of the cosmic microwave background radiation then confirmed the reality of the Big Bang. So today we refer to the Big Bang theory (because it has passed the tests), whereas the Steady State is still just a model.

1. In an open universe, the angles of a triangle add up to less than 180°.

2. In a flat universe, the angles of a triangle add up to exactly 180°.

3. In a closed universe, the angles of a triangle add up to more than 180°.

Einstein's Geometry

The general theory of relativity describes gravity in terms of bent spacetime. Around a black hole, the influence of the matter in the black hole bends spacetime round upon itself so that nothing can escape. The space around a black hole is like the surface of a sphere (or the surface of the Earth), and is said to be closed. Just as if you keep going in the same direction on Earth you go round the planet and back to where you started, so in closed space if you keep going in a straight line you go round the universe and back to where you started. The inside of a black hole really is a closed universe. Nothing can escape.

At the other extreme, gravity can bend space in the opposite sense. This is hard to picture, but an analogy would be the surface of a saddle, or a mountain pass, which curves away in all directions. Such a surface is said to be open. A closed universe is one which cannot escape from its own gravitational grip; an open universe is one which is expanding faster than its own escape velocity.

But there is a special case in which space is flat, like the surface of a smooth desktop. In Einstein's geometry, this is the equivalent of the special case where the ball is travelling upwards from the Earth at exactly the escape velocity. It is the only unique kind of geometry allowed by the general theory of relativity – there are lots of open universes, and lots of closed universes, but only one flat universe. And the real Universe seems to have a geometry indistinguishable from this very special case.

Close to Critical

Because the flat universe model is unique, cosmologists use it as a benchmark against which to measure other models. The flat universe model is said to have 'critical density', which means that the density of the universe is exactly right to make space flat. Cosmologists measure the density of the Universe (or model universes) in terms of a parameter called omega (Ω). It is also known as the flatness parameter. For a flat universe $\Omega = 1$. For an open universe, Ω is less than 1, and for a closed universe, Ω is bigger than 1.

The Universe we live in is expanding, and that means that the density is decreasing as time passes. This affects the value of Ω at any moment of cosmic time (any epoch), but it does not affect which side of the critical dividing line Ω sits. Leaving aside some of the more exotic complications that can be caused by a cosmological constant, if the Universe was dense enough to be closed when it

Alexander Friedmann's application of the general theory of relativity produced a variety of model universes in the early 1920s – before astronomers knew for sure that there was a Universe beyond the Milky Way.

1

emerged from the Big Bang, then it will stay dense enough to be closed. And if it started out open, then it will always be open.

Going to Extremes

Whichever way the Universe started out, the fact is that it shifts further and further from that critical density as time passes. If it started out with Ω just less than 1, as it expands and the density decreases, the value of Ω gets less and less. If i t started out with Ω just bigger than 1, the expansion proceeds more slowly and the value of Ω gets bigger and bigger as time passes. By now, some 14 billion years after the Big Bang, there has been ample time for this effect to have been at work. The Universe should have evolved to one extreme or the other. But when we look at the real Universe, we see that it is still very close to the critical density.

It is hard to measure the density of the entire Universe, but astronomers can make a good stab at it. First, they count all the bright galaxies in a chosen volume of space, and estimate the mass of all the bright stars in those galaxies. Then, from the way galaxies move in clusters under the influence of gravity they can estimate how much 'dark stuff', or 'dark matter'(▷ p. 117), there is, tugging on the bright stuff gravitationally, for every galaxy they see. Adding all this up, they find that there is matter to account for at least 10 percent of the critical density of matter around, and probably at least 30 percent. So, based on the dynamics of galaxies, Ω is at least 0.1 and may be bigger than 0.3. But for 14 billion years, Ω has been getting further and further away from 1. For Ω to be as big as 0.1 today means that in the first second of the Big Bang it must have been within one part in 10^{60} (a 1 followed by 60 zeroes) of being precisely 1. This means that the Universe was born flat to an accuracy of 1 part in 10^{60}. The flatness of the Universe in the Big Bang is the most accurately determined parameter in the whole of science.

THE ARROW OF TIME

One of the most puzzling things in science is the origin of the 'arrow of time'. We all know that there is a difference between the past and the future, but where does this difference come from? There is nothing in the laws of physics which makes the distinction. In the classic example of two billiard balls colliding and moving apart, the laws of physics 'work' just as well to describe the same collision with time running backwards. However, the arrow of time seems to have something to do with the way very large numbers of things interact with one another. In the break of snooker, for example, it is quite clear in which direction time goes.

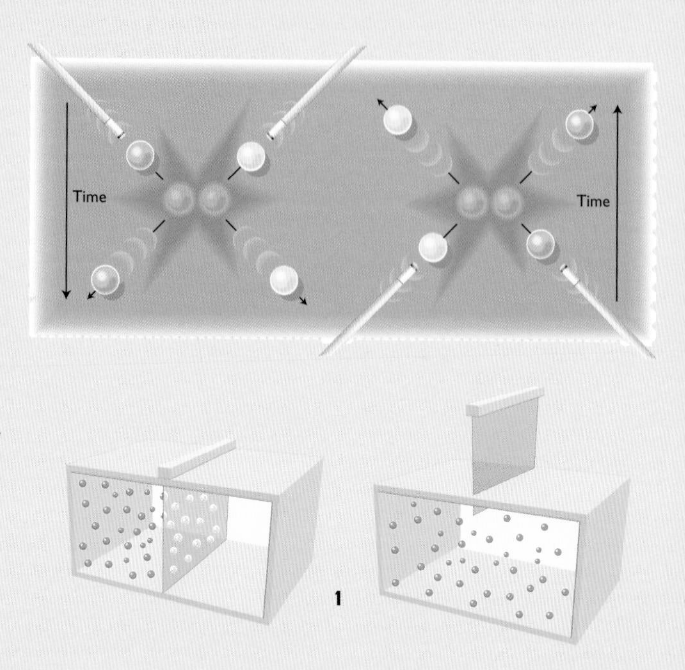

1

Time in a Box

Think of a simple box, divided in two by a sliding partition, with smoke in one half and empty space in the other. When you slide open the partition, the smoke spreads out to fill up the box. However, no matter how long you wait, you will never see the smoke move back into one half of the box so that you can slide back the partition and trap it. If you saw two pictures of the box, one with smoke in one half of the box and one with smoke in all of the box, you would know which picture was taken first.

Mathematically, the difference between the two situations is that you need less information to describe the full box. There is a pattern (or order) in the half-empty box (half-empty, half-full) that is lost when the box is uniformly full of smoke. Information, or order, is measured by a quantity called entropy, in such a way that a decrease in information (increase in disorder) corresponds to an increase in entropy. Overall, in the Universe at large, entropy always increases as time passes.

2

3

Local Order

Left to their own devices, things wear out. Cars rust, glasses break, houses fall down. Disorder increases. We can only make order locally (making cars, building houses and so on) using energy. On Earth, the energy comes, ultimately, from sunlight. But the increase in entropy, caused by the processes inside the Sun which release energy, is much bigger than the decrease in entropy caused by the action of life on Earth. In the Universe at large, entropy always increases, even though it may temporarily decrease on a planet like Earth. Time points in the direction of increasing entropy.

4

5

1. Although individual collisions between atoms and molecules seem to be reversible, the behaviour of lots of molecules in a gas reveals the direction of time.

2 and 3. Rusting cars (left) and burning fires (4 and 5 above) both indicate the arrow of time in the real world.

Time and the Universe

The Universe was born in a state of low entropy. Processes going on inside the Sun and stars increase the entropy of the Universe as energy pours out into the cold of space.

Another indication of the arrow of time is that in nature heat always flows from a hotter object to a cooler object. So another way of defining the arrow of time is that it points away from the hot Big Bang and into a cold future. When all the stars have burned out, everything in the Universe will be at the same temperature. It will be uniform, with no pattern or order, and there will be no way of telling one locality from another. Time will come to an end. This is called the 'heat death' of the Universe and it is inevitable if the Universe is destined to expand forever. The best evidence today is that this will be the fate of our Universe.

TOPIC LINKS

2.1 The Big Bang
p. 87 The Singular Beginning

2.2 Cosmology for beginners
p. 100 Reversing Time

2.4 The Accelerating Universe
p. 129 The Fate of the Universe

4.2 A Choice of Universes
p. 209 Hawking's Universe

2.3 MISSING MASS AND THE BIRTH OF TIME

After the 1960s, cosmologists began to accept that the Big Bang idea was more than just a model and offered a good description of the real Universe. But as it stood at the time, this was far from being an exact description of the Universe. The basic idea of the Big Bang seemed sound, but there were problems relating the detailed appearance of the Universe we live in to the physics of the Big Bang itself. These puzzles centre around the observational evidence that the Universe we live in is uniform but nevertheless contains irregularities.

A breakthrough came when particle physicists started to apply their ideas, based on studies of what happens to particles at very high energies, to the physics of the Big Bang. The resolution of those puzzles involved new ideas about what happened in the earliest moments of the existence of the Universe, and new observations to test those ideas. The result was a golden age of cosmology, which is still going on today.

Previous page. Optical image of the Omega Nebula 6000 light years away from Earth in the constellation Sagittarius.

BIG BANG PROBLEMS

Before about 1960, cosmology was still a mathematical game, in which a few experts (probably no more than 20 in the whole world) explored the possible model universes allowed by the equations of the general theory of relativity. In the 1960s, they were delighted, but rather surprised, to find that one of those models, the Big Bang, actually seemed to provide a very good description of the Universe we live in. The discovery of the background cosmic radiation, and the elucidation of how the primordial hydrogen and helium that went into the first generation of stars in the first four minutes of the Big Bang was made, made people start to take the Big Bang model seriously. But in the 1970s, cosmologists (by now there were hundreds of them) began to realize that there was a problem they had never expected with the Big Bang theory: it was, in a sense, too good to be true.

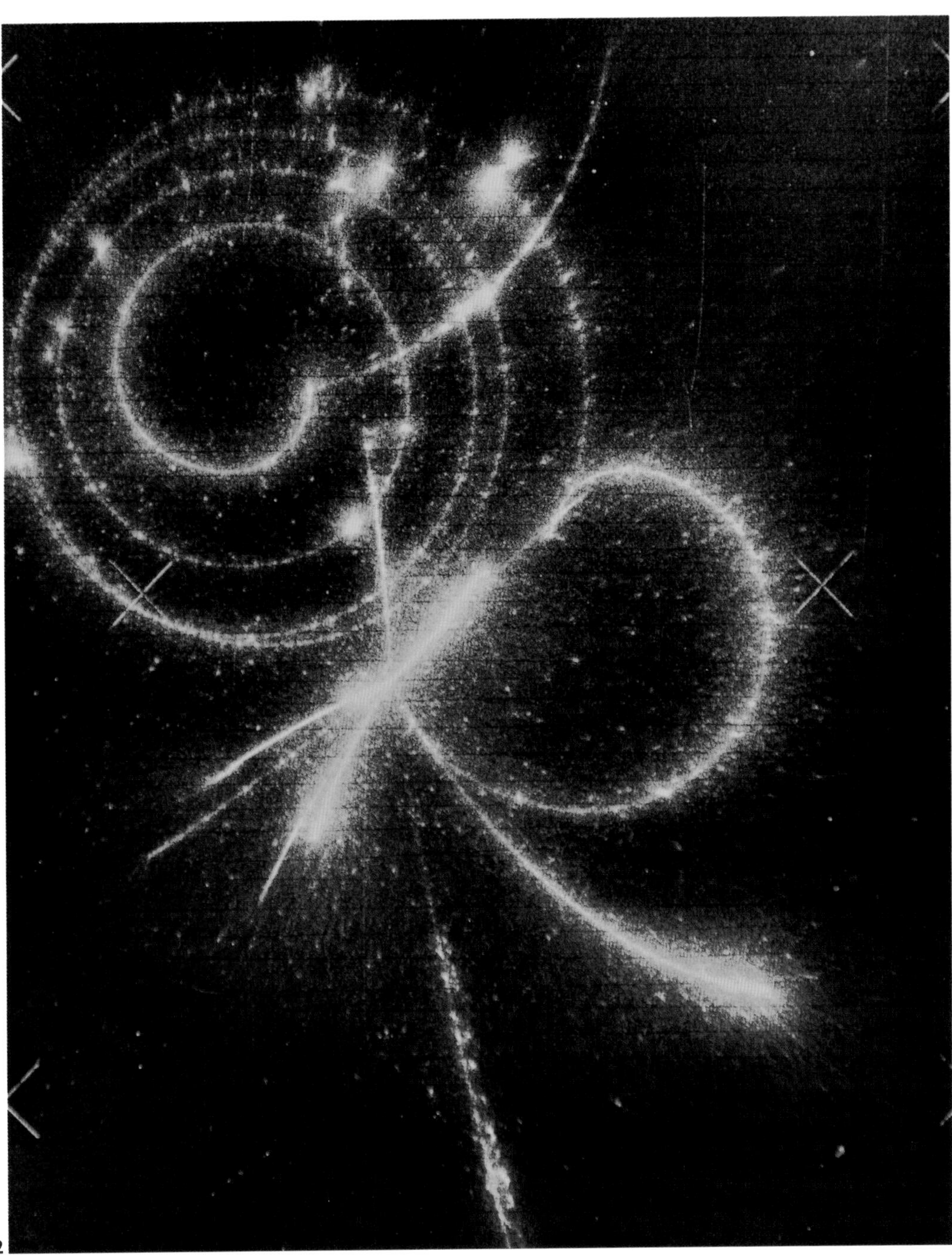

1 and 2. Photographs showing the interactions between sub-atomic particles in experiments at the high energy accelerators of CERN at Geneva.

It took only 0.000000000001 s (10^{-12}s) for what was to become the entire visible Universe to expand from a size smaller than an atom to the size of the Solar System.

1. On the scale of stars and galaxies, the universe doesn't look smooth. Stars form patterns like the southern cross.

2. Flatness can be a matter of perspective – the ground beneath our feet seems flat even though the Earth is round.

The Flatness Problem

In fact, there were several 'too good to be true' problems that puzzled cosmologists in the 1970s. The first is the question of why the Universe is so nearly (perhaps precisely) flat (▷ p. 95). As we have seen, whichever side of the critical density it started out from in the Big Bang, the Universe should have got further and further away from $\Omega = 1$ as time passed. For the Universe still to be so nearly flat after nearly 15 billion years is about as likely as balancing a pencil very carefully on its point on a table, then going away for 15 billion years and coming back to find that the pencil is still as you left it. The only explanation seemed to be that there ought to be a law of nature which forced the Universe to be precisely flat. But, in the 1970s, nobody could think what that law might be.

2

The Smoothness Problem

Another big puzzle about the Universe is that it is incredibly smooth. You might not think so, looking at the patterns made by stars on the night sky, or the way galaxies are grouped in clusters across the dark expanse of the Universe. But these are really small-scale effects. What matters is the smoothness of the dark space itself – in Einstein's language, the smoothness of spacetime. From a distance, the surface of the Earth looks smooth and, compared with its diameter, it is. A mountain 12.7 km high is only a pimple corresponding to one-thousandth of the diameter of the planet, even though it looks enormous to us. In the same way, the uniformity of the Universe has to be measured on the largest possible scale, and that means using the cosmic background radiation – the echo of the Big Bang.

The background radiation has the same temperature from all parts of the sky – it is

Although nothing can travel through space faster than light, space itself can expand faster than light.

1

2

isotropic. This shows that the Universe emerged from the fireball of the Big Bang in a very smooth state. We have to get used to the idea that even clusters of galaxies are like tiny pimples on the smoothness of the Universe.

There's another way of looking at this. The Universe isn't just the same in all directions, it is the same (on average) *everywhere*, once you allow for the effects of expansion. The pimples are distributed at random across the face of space. The buzzword this time is 'homogeneity'. You can see the homogeneity of space by looking at two stunning pictures from the Hubble Space Telescope, called Deep Field North and Deep Field South. These are images of very distant galaxies seen on opposite sides of the sky. But apart from superficial differences, they look exactly the same. The same kinds of galaxies are clustered together in the same kind of way. There is nothing in the pictures to tell you which part of the Universe those galaxies inhabit, although they occupy regions of space billions of light years apart (just as there is nothing in a picture of a pimple to tell you which cheek of an acne sufferer it afflicts). There are no 'lumps' in the Universe, except for the galaxies and clusters of galaxies themselves. It is homogeneous. Or nearly; there are, after all, galaxies. And in the 1970s, this was another problem.

The Problem of Galaxies

As observations of the background radiation got better and its smoothness was measured with increasing precision, it began to be difficult to see how galaxies could have formed at all. Galaxies start out when huge clouds of gas collapse under their own gravitational pull, their own weight. But this collapse can only

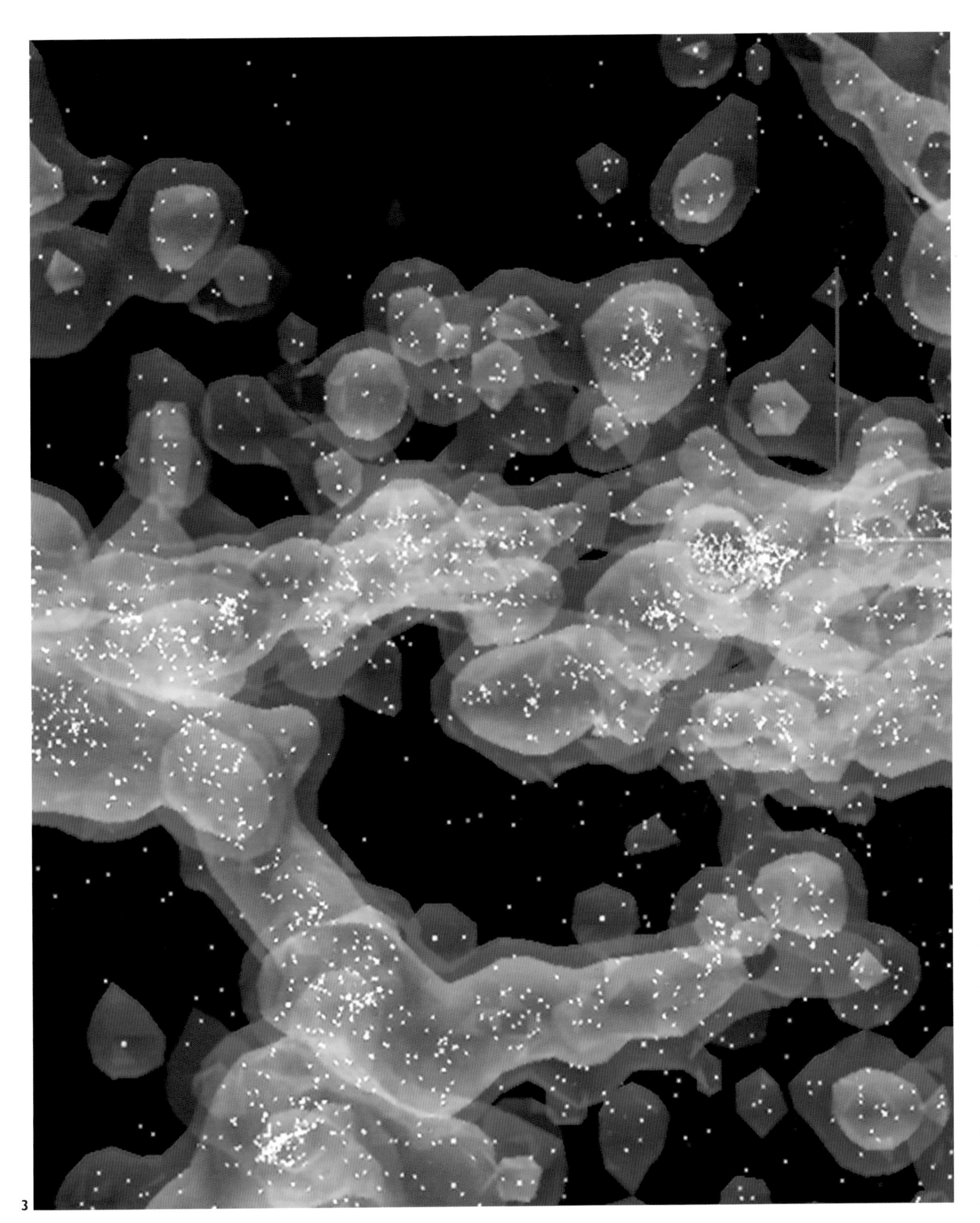

3

1 and 2. The universe looks the same in all directions. The top image is a picture obtained by the Hubble Space Telescope looking north on the sky. The bottom picture is a similar image obtained looking south.

3. The distribution of galaxies on the sky is not entirely random, but forms clumps and filaments.

1. The microwaves used to map the Universe are similar to the ones used in telecommunications.

2. Alan Guth, one of the pioneers of the theory of inflation.

3. (opposite) Computer simulation of the way matter clumps together in the expanding Universe.

begin if some regions of the Universe are more dense than others. If everything were perfectly smooth, gravity would pull evenly in all directions, and everything would get carried along with the expansion of the Universe. Like seeding clouds to make rain, you need some initial lumpiness to provide the seeds on which galaxies can grow. These seeds must have been present at the end of the fireball era of the Big Bang, 300,000–500,000 years after the moment when time began, at the time when the radiation that became the microwave background last interacted with matter.

So the unevenness in the Universe from that time, that led to the formation of galaxies, should have left its imprint on the microwave radiation. Why hadn't this been seen?

The answer was that it was too small. When cosmologists carried the calculations through, they found that the seeds from which galaxies and clusters of galaxies formed should indeed have left a mark on the background radiation, but a mark corresponding to differences in temperature today from one part of the sky to another of only about 30 millionths of a degree. Nothing highlights more clearly just how insignificant galaxies (let alone stars, planets and people) are to the cosmos at large. There was no hope (at that time) of detecting, from the ground, such tiny ripples in the background radiation. In the mid-1970s, a group of astronomers took up the challenge of designing and building a satellite which could look for the predicted disturbances in the background radiation. The result, COBE, did not fly until the end of the 1980s, when it triumphantly confirmed the predictions of the theorists. But by then, the theory of the Universe had itself undergone a radical rethinking that solved the problems of the Big Bang.

Flattening the Universe

The standard model of the Big Bang describes everything that happened in the Universe from about 0.0001 s (10^{-4}s) after 'time zero', when the temperature of the Universe was 1000 billion degrees (10^{12} K), up to the time about half a million years later when matter and radiation decoupled at a temperature of 6000 K. It never pretended to explain what went before that, when temperatures were higher than 10^{12} K. The understanding of physics that existed in the 1960s was inadequate to describe what went on that close to the singularity. But as physicists on Earth

of galaxies and clusters of galaxies that we see around us could have formed in the time available since the Big Bang. Adding enough HDM to flatten the Universe makes things even worse. But if there is at least 30 per cent of the critical density in the form of CDM, then the pattern of galaxies and clusters that emerges from the simulations is strikingly similar to the pattern we see in the real Universe. This is powerful evidence that CDM is the dominant form of matter in the Universe, although these results do allow for the possibility that there is a little HDM as well (perhaps more HDM than there is baryonic matter).

This picture of a Universe dominated materially by CDM was firmly established by the end of the 1990s. But it still isn't quite the last word. In order to make galaxies and clusters that look like those in the real Universe, you need less than half of the critical density in the form of dark matter – but the models still work best if that matter is embedded in flat space. If the density of matter is less than half the critical density, how can space be flat? The answer takes us back to Albert Einstein's original thoughts on cosmology, but also right up to date with key results from the astronomical frontier announced at the beginning of the twenty-first century, seeming to slot the last piece into the cosmological jigsaw puzzle.

The surprising implication of this work, yet to be confirmed, is that the Universe may be precisely flat, but it is also expanding faster as time goes by.

1. (opposite) Computer model of dark matter (red) distribution, the first time the invisible dark matter has been mapped.

2. Experiments designed to detect dark matter particles have to be buried deep underground to minimise interference.

WHY THE SKY IS DARK AT NIGHT

You can make one of the most profound observations in the whole of science by going outside and looking at the dark night sky. Why is the sky dark at night? Is it because there are gaps between the stars? But if space went on to infinity, and it was filled with stars, every 'line of sight' out into space would end on a star, and the whole sky would blaze with light. So the Universe cannot be infinite and uniform.

1

Olbers and the Edge

The puzzle of the dark night sky often goes by the name 'Olbers' paradox', after the nineteenth-century German astronomer who publicised it. But he wasn't the first to ponder the puzzle, and it isn't really a paradox. The simplest way to picture the puzzle is to imagine standing in a large forest. Wherever you look, you will see a tree. But if you are only standing in a small wood, you might be able to see through the spaces between the trees, out to the edge of the forest. In an infinite Universe, everywhere you look you will see a star. The fact that we can see 'out' through the gaps between the stars seemed to imply, to Olbers and others, that there must be an 'edge' to the Universe, beyond which there was only dark, empty space, and no more stars. This reasoning applies even if you think in terms of galaxies rather than stars.

Looking Back in Time

The first person who realized that this need not be the case was Edgar Allan Poe who, alongside his literary activities, was a keen amateur scientist. He gave a lecture setting out the correct resolution to Olbers' paradox in February 1848. But he died a year later and no scientists took up his suggestion.

What Poe pointed out was that by looking further out into space we are looking further back in time, because it takes light a finite time to travel through space. When we look through the gaps between the stars at the dark night sky, we are looking back in time to an era before the stars were born – in Poe's words, to a distance 'so immense that no ray from it has yet been able to reach us at all.'

The Edge of Time

This notion needs only a tiny modification to fit in with the idea of the Big Bang. The point is that the Universe is not infinitely old, and it has an 'edge' in time (the Big Bang) rather than an edge in space. Looking through the gaps between the stars and galaxies, we do indeed look back to a time before galaxies formed.

But modern instruments reveal that the sky is not completely dark. What we 'see' there is the background radiation from the Big Bang fireball itself, once as hot as the surface of a star, but now redshifted by the expansion of the Universe down to 2.7 K.

The darkness of the night sky is evidence that the Universe was born at a definite moment in time. You can see (or rather, not see!) evidence to support the Big Bang model with your own unaided eyes.

existence out of nothing at all, provided they annihilate one another and disappear within a certain time. The time limit is set by their mass – the bigger the mass, the smaller the time they can exist for – but is measured in tiny fractions of a second. It is as if the particles exist when the Universe isn't looking, but as soon as it has time to notice their presence, they vanish.

The effect of all this is to make space a seething foam of virtual particles, giving it both energy and structure. It is the energy that provides the outward push of the term, and the mass associated with that energy that completes the job of making the Universe flat.

It is no coincidence, by the way, that all the different kinds of mass-energy in the Universe add up to make it flat, with $\Omega = 1$. Inflation drives the Universe to flatness, so there is only so much mass-energy to go round, to be shared among baryons, HDM, CDM and Λ. It's as if you had a variety of different bottles and jars and you poured water into them from a one-litre container. No matter how the water is divided among the different containers, the total amount of liquid has to add up to one litre.

But although quantum fluctuations are usually ephemeral, it seems they have left their imprint on the Universe.

1. In a fractal, each small piece of the pattern can be enlarged to reproduce the whole pattern.

A Matter of Scale

As well as occurring in tiny split-seconds of time, quantum fluctuations also occur on tiny distance scales (not least because the disturbances involved do not have time to travel far before they are forced to disappear). These processes were going on in the earliest phase of the Universe, after the Planck time and before inflation. At the time inflation took a grip on what was to become the visible Universe, all of the mass-energy associated with the visible Universe was tucked inside a tiny seed just 10^{-25} cm across – 100 million times bigger than the Planck length but still 1000 billion times smaller than a proton. Even this ridiculously tiny seed was big enough to contain quantum fluctuations, involving energetic fields (like electromagnetism) rather than particles. So the vacuum had an ever-changing structure, but the structure always matched a certain statistical pattern.

Then inflation happened. Everything in the universal seed was ripped apart and spread out. In the process, whatever fluctuations of the vacuum were going on at the moment inflation began were frozen into the structure of the rapidly expanding seed and also enormously stretched out as space expanded. Space actually expanded faster than light during inflation (this is entirely allowed by Einstein's equations; it is only motion *through* space that cannot exceed the speed of light), and the last quantum fluctuations were imprinted on the pattern of hot gas that emerged from the cosmic fireball.

The statistical pattern of the quantum fluctuations is called 'scale invariance', because it looks the same (statistically) on all scales – if you take a piece of the picture and expand it, it doesn't look precisely like the original, but has the same statistical appearance, in terms of the arrangement of hot spots and cold spots. What COBE and its successors see in the ripples in the background radiation is exactly the same pattern of scale invariance, but 'written' over hundreds of millions of light years, instead of within a sphere 1000 billion times smaller than a proton. We are part of that pattern – life is part of the structure imprinted on the Universe by quantum fluctuations shortly after the birth of time.

BALLOONS AND THE BACKGROUND RADIATION

Technology has improved so much since the time of COBE that, in the absence of any new microwave satellites, the best maps of the microwave background radiation now come from instruments attached to balloons on flights above most of the Earth's atmosphere. Until the next generation of microwave satellites goes up, taking comparable instruments with them right into orbit, the best picture of the microwave Universe comes from one of those experiments, Boomerang.

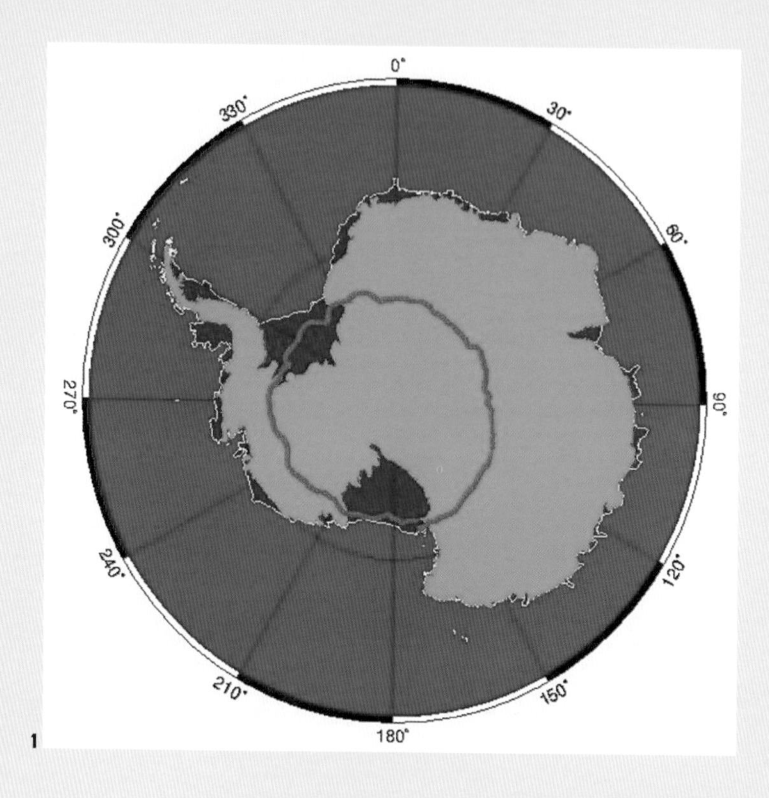

1

2

Around the World in 10.5 Days

Boomerang gets its name from the fact that the balloon travels in a roughly circular path around Antarctica, carried by high altitude winds. Technically, because it circles the South Pole, this is a round-the-world flight. Its first circumpolar flight started from the McMurdo base at 03:30 GMT on 29 December 1998, and ended when the balloon returned to its starting point at 15:50 GMT on 8 January 1999. It went around the world in 10.5 days.

The microwave telescope used to monitor the background radiation was lifted to an altitude of 40 km by a balloon the size of an American football stadium.

Why Antarctica?

Antarctica is a very good place to operate a balloon mission like Boomerang for several reasons. First, because the path of the balloon is predictable and it returns to its starting point, it can stay aloft for a long time. It cannot stay up as long as a satellite, but similar balloons launched in other parts of the world have to be brought down in a few hours (at most a couple of days) before they become a hazard or get lost.

Secondly, the air above Antarctica is cold and dry, which means that even the remaining trace of atmosphere above 40 km altitude does not have much influence on the incoming background radiation, which can be partly absorbed by water vapour.

Finally, because Antarctica is uninhabited (except by scientists and penguins) the balloon is not going to get in anyone's way, and there are no local radio and television stations to interfere with its microwave detectors.

Supercool Science

Even with all these advantages, mapping the

microwave sky accurately is difficult. Basically, the detectors are measuring the temperature of the background radiation from different parts of the sky, and they can only do this accurately if the detectors are even colder than the radiation (the colder the better). The radiation has a temperature of 2.735 K. The Boomerang detectors were cooled to 0.28 K (–272.88°C, since 0 K is –273.16 degrees Celsius) in a giant Dewar (like a Thermos flask) at the focus of a 1.3 metre diameter telescope.

The Boomerang Dewar contained 65 litres of liquid helium in an inner container and 75 litres of liquid nitrogen in an outer container. Together these are sufficient to keep the detectors at the required temperature for up to 12 days.

Confirmation by Balloon

Although other launch sites are less ideal than Antarctica, other balloon experiments provided valuable confirmation that Boomerang really is detecting fluctuations in the background radiation, and not some spurious 'signals' caused by problems with its detectors. The most important of these complementary balloon programmes was dubbed MAXIMA, from Millimeter Anisotropy eXperiment Imaging Array.

The first flight of MAXIMA took place in August 1998, and lasted for just 4.5 hours. It uses a similar 1.3 metre telescope to the one on Boomerang, cooled to about 0.1 K. On this and subsequent flights, MAXIMA found the same kind of fluctuations as Boomerang, but in the northern sky (it was launched in Texas) not the southern sky. Although the results were not as impressive in themselves as those from Boomerang, they were crucially important because they showed the same pattern from different parts of the sky, confirming that what Boomerang saw is a universal effect.

3

1. Map of the flightpath of the Boomerang.

2. Boomerang's map of the microwave sky. The colour variarions represent varaitions in temperature.

3. The launch of MAXIMA in August 1998.

TOPIC LINKS

2.1 The Big Bang
p. 83 Microwaves from the Birth of Time
p. 84 The Very First Light
pp. 90–1 The First Four Minutes

2.3 Missing Mass & the Birth of Time
p. 116 Ripples in the Background Radiation

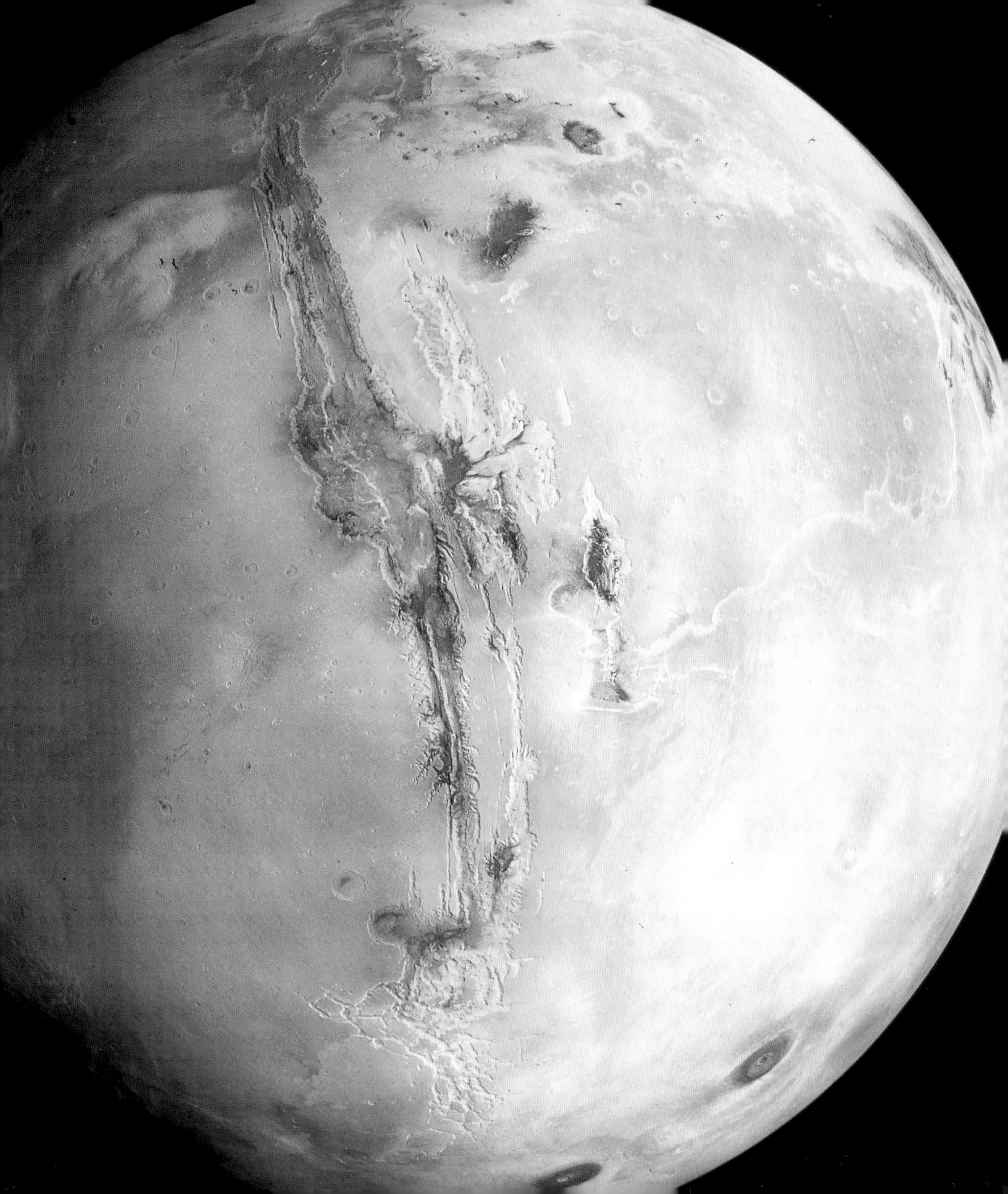

1

chance of finding more than one planet in the equivalent comfort zone around another star, and all kinds of things might go wrong to prevent life getting a grip on that one suitable planet. But when spaceprobes visited the outer part of our own Solar System, they found that this pessimistic point of view might be wrong.

Extending the Comfort Zone

Assuming that life elsewhere resembles life on Earth (which is the only basis we have to go on), the search for life is essentially the search for liquid water. Without liquid water, there is no life. It is no coincidence that our word for a region on Earth devoid of life, desert, is the same as our word for a region devoid of liquid water.

At first sight, it looks as if the surface of any planet beyond the orbit of Mars should be too cold for liquid water to exist. But in the late 1990s, the space probe Galileo sent back pictures from one of Jupiter's moons, Europa, which revealed that the moon is almost entirely covered by what seems to be a layer of ice floating on an ocean of liquid water, very like pack ice floating on the Arctic Ocean. Although it is smaller than our own Moon, Europa still has a diameter of 3,138 km, making it a respectably sized potential home for life. But what keeps it warm?

The answer seems to be that as Europa follows its orbit around Jupiter, the gravitational pull from Jupiter itself, and from the other moons orbiting around the planet, combine to produce a changing tidal force, constantly 'kneading' the inside of Europa. It is squeezed rhythmically in and out, in the same way that the gravitational pulls of the Sun and Moon cause the tides to move rhythmically up and down the seashore on Earth. This kneading generates heat, enough heat to melt the ice which makes up the bulk

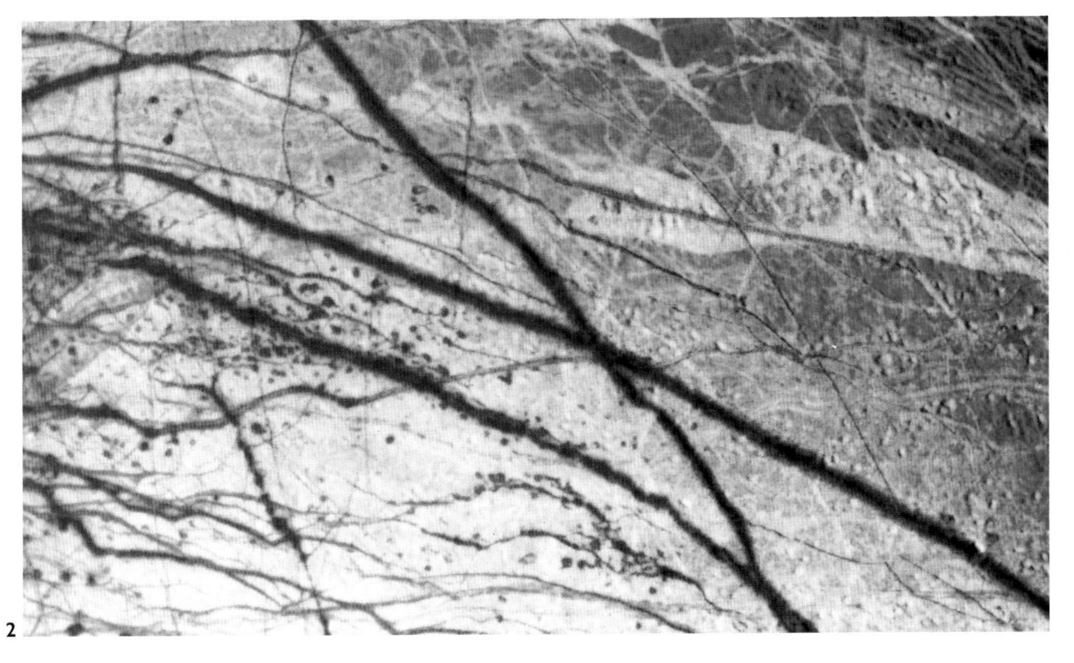

When pulsars were first discovered, it was thought that their regularly repeating 'signals' might be messages from little green men!

1. Family portrait of Jupiter and its four largest moons.

2. Close-up view of the icy surface of Europa, one of Jupiter's moons.

3. The life zone around the Sun.

of Europa. The result might not be comfortable for human beings, but if life can exist in the chilly waters of Antarctica, then in principle it can exist on Europa.

At once, the known comfort zone for life around our Sun doubled in size.

Heat from Inside

There are other ways to generate heat. The Earth itself is hot inside because of radioactivity, which keeps the core molten. If a planet like the Earth formed in the equivalent of an orbit between those of Jupiter and Mars, it might be too cold on its surface for liquid water to flow, but it could have very deep, frozen oceans, because very little of its primordial water would have evaporated. Any ice deeper than 14 kilometres below the surface would be melted by the internal heat of the planet.

At the end of the twentieth century, astronomers realized that they had been far too parochial in their estimates of the chance of finding liquid water elsewhere in the Universe and, therefore, in their estimates of the chance of finding life. But the Holy Grail still remains: what are the chances of finding other Earths?

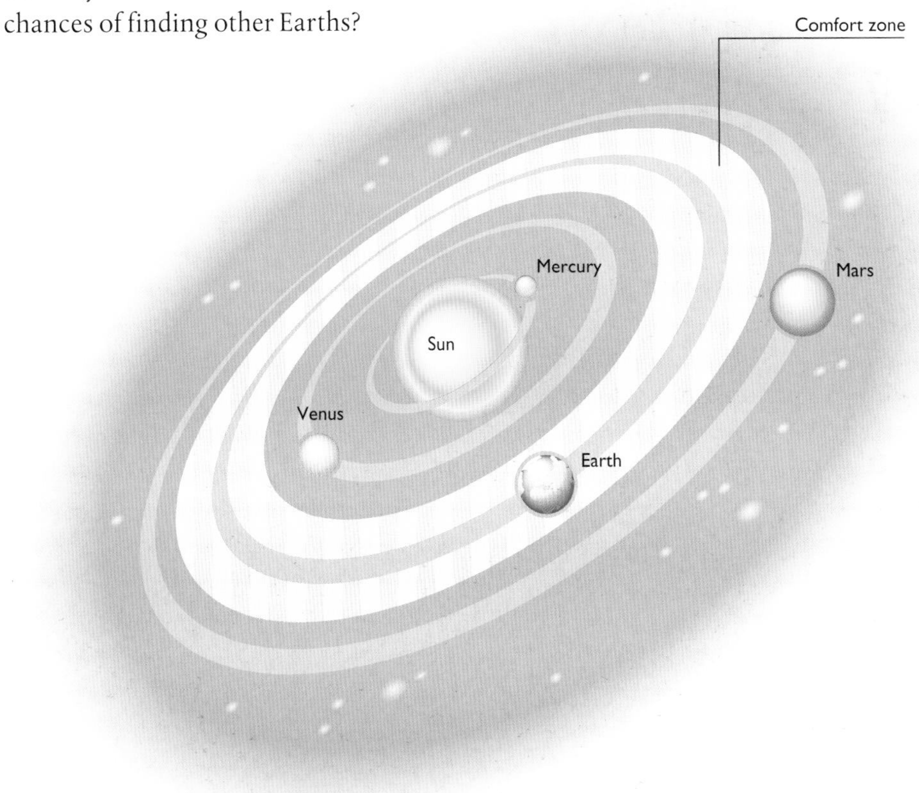

LIFE ON EARTH

One of the biggest mysteries in biology is how life got a grip on the Earth so soon after our planet formed. The Earth formed about 4.5 billion years ago, but it was bombarded with debris from space early in its life, and took about half a billion years to cool to the point where liquid water could flow. Yet there is fossil evidence of bacterial life in rocks at least 3.9 billion years old. Could life really have got started from scratch in just a hundred million years or so?

2

Cosmic Rain

Perhaps life didn't have to start from scratch on Earth. Although the bombardment of the Earth eased after about half a billion years, it didn't stop. Even today, the Earth is still occasionally struck by a large object (as the dinosaurs found to their cost 65 million years ago), and more gentle impacts with comets and smaller bits of cosmic debris happen quite often – on a geological timescale, that is. Since the 1960s, there has been a growing awareness that clouds of material in interstellar space contain a variety of complex carbon compounds, the raw materials of life. It now seems likely that cometary impacts when the Earth was young brought some of those raw materials down to the surface of our planet, giving life a kick start.

In 1986, cameras onboard the space probes Giotto and Vega sent back images of dark material coating the surface of the icy core of Halley's comet, and spectroscopy revealed that this is made of a variety of carbon-rich molecules. Ground-based telescopes have shown that the gases which make comets so spectacularly visible are also rich in carbon compounds, including methane and ethane. All this is important, because carbon is the key component of life (so much so that complex carbon chemistry is also called 'organic chemistry'). Microscopic particles of dust from space, much of it comet debris, are constantly falling onto the Earth, and analysis of samples collected at high altitudes tells us that about 30 tonnes of organic carbon compounds reach the surface of the Earth this way each day.

1

Seeds from the Clouds

The ultimate origin of all this stuff is the interstellar molecular cloud in which our Solar System was born. Spectroscopic studies have revealed the presence of many organic carbon compounds, such as formaldehyde, and ethanol (also known as vodka). But the most important discovery, made in 1994, was that of glycine, the first amino acid discovered in space. Out of more than a hundred molecules known to exist in space, this is the most important, because amino acids are the building blocks of proteins, and proteins are what your body is made of.

It is very hard to see how simple compounds like carbon dioxide and water could have developed into living bacteria in only a hundred million years. But if the young Earth were laced with complex organic molecules things like amino acids, then the whole process would have proceeded much faster.

The problem of the origin of complex organic molecules, the precursors of life, is moved from the Earth and out into space. There, although cold clouds of gas and dust may not seem the ideal places for chemical processes to take place, there was time enough – billions of years – for these first steps to be taken and for the seeds of life to form.

This would have happened before the stars and planets formed from the collapse of those clouds. The implication is that the 'seeds' of life fall upon every new planet, even if they do not always 'germinate' there.

SIGNALLING TO THE UNIVERSE

But how can we hope to communicate meaningfully with alien civilizations, rather than just making a noise? The scientists involved in sending the 1974 signal from Arecibo wanted to send a message that contained information that an alien might be able to decipher, so they used what is considered the universal language – mathematics. One of the scientists involved, Carl Sagan, described the message in words at the time:

'What it said fundamentally was: "Here's the Sun. The Sun has planets. This is the third planet. We come from the third planet. Who are we? Here is a stick diagram of what we look like, how tall we are, and something about what we're made of. There's four point something billion of us, and this message is sent to you courtesy of the Arecibo telescope, 305 metres in diameter."'

It turns out to be surprisingly easy to convey that kind of information in the simplest mathematical language of all – the binary 'on-off' code familiar from computers. Frank Drake, a keen enthusiast for SETI (see later), tested this in the 1960s by devising just such a message (later used as the basis for the Arecibo message) and sending it out to colleagues to see if they could crack the code.

Drake's Message

Drake's message consisted simply of a string of 0s and 1s – binary code – 551 characters (or 551 'bits' in computer language) long. Any mathematician would quickly realize that

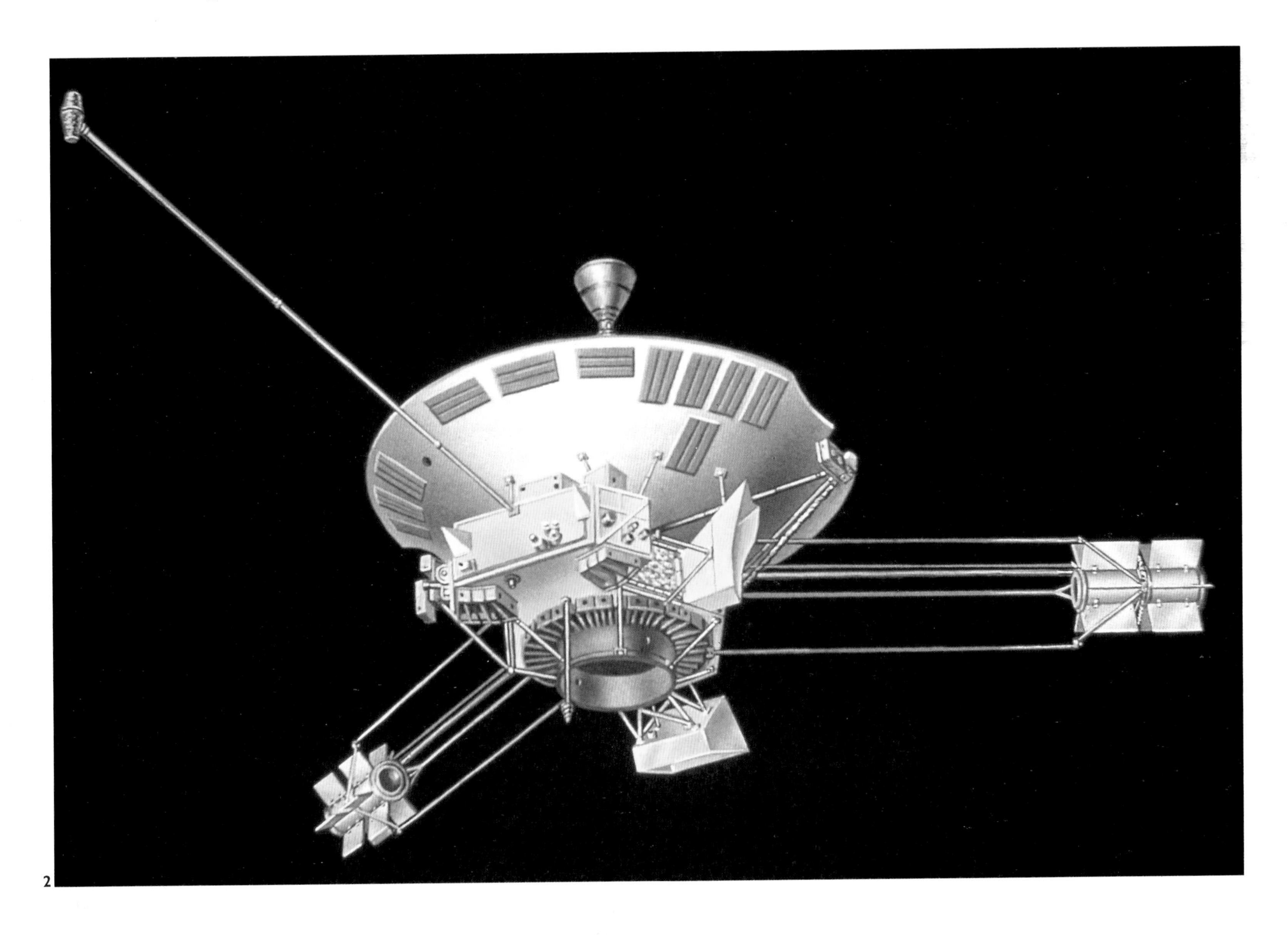
2

> The name 'nano', as in 'nanobacteria' and 'nanometre', comes from the Greek word *nannos*, meaning a mischievous dwarf.

551 is a product of two prime numbers, 19 and 29. The only two numbers that multiply to give 551 are 19 and 29. But 19 and 29 do not divide by any whole number. This suggests (to the mathematically inclined) that the string of 0s and 1s can be turned into a rectangular 'picture' in either of two ways – a grid with 19 rows of 29 characters, or a grid with 29 rows of 19 characters. Trial and error would quickly show that the first grid is gibberish, but the second produces a distinct pattern if you put a black square everywhere there is a 1 and a white square where there is a 0 (or vice versa).

Drake made up his message as if it came from an alien civilization, so the pattern he made represents the imaginary civilization. The little picture of the alien itself is fairly obvious. Down the left-hand side of the picture there is a representation of a star and its nine planets, with a rough indication of their sizes. The other bits of code are mostly numbers. The numbers 1 to 5, written in binary, alongside the first five planets, then 11 alongside planet two, 3000 alongside planet three, and 7 billion, alongside planet four. Drake intended this to mean that 7 billion aliens lived on planet four, they had a colony on planet three, and that at the time they sent the message there was a small expedition on planet two. The diagrams at the top right of the picture represent the atoms carbon and oxygen, telling us a little about the aliens' chemistry.

No individual scientist managed to decipher all of Drake's 'alien message' (by the way, he didn't pretend it really did come from aliens; they knew he had made it up as a test). But any real message from the stars would be investigated intensively by high-powered teams of scientists from all disciplines and they would be able to crack any similar code. This shows what an impressive amount of information can be packed into just 551 bits of computer code. With 8 bits making a byte, that is less than 70 bytes of information, and computer memory these days is measured in megabytes and gigabytes.

1

Messages in Cosmic Bottles

You may think that radio waves take a long time to get to the stars, but space probes, crawling along at a tiny fraction of the speed of light, take vastly longer. Even so, space scientists have already sent messages beyond the Solar System attached to space probes. These were not probes designed for the purpose, but ones that happened to be heading out of the Solar System anyway, after completing their missions of exploration of the outer part of the Solar System. This is the equivalent of throwing a message in a bottle into the ocean, and hoping that it might be picked up by a passing ship; a long shot, but the

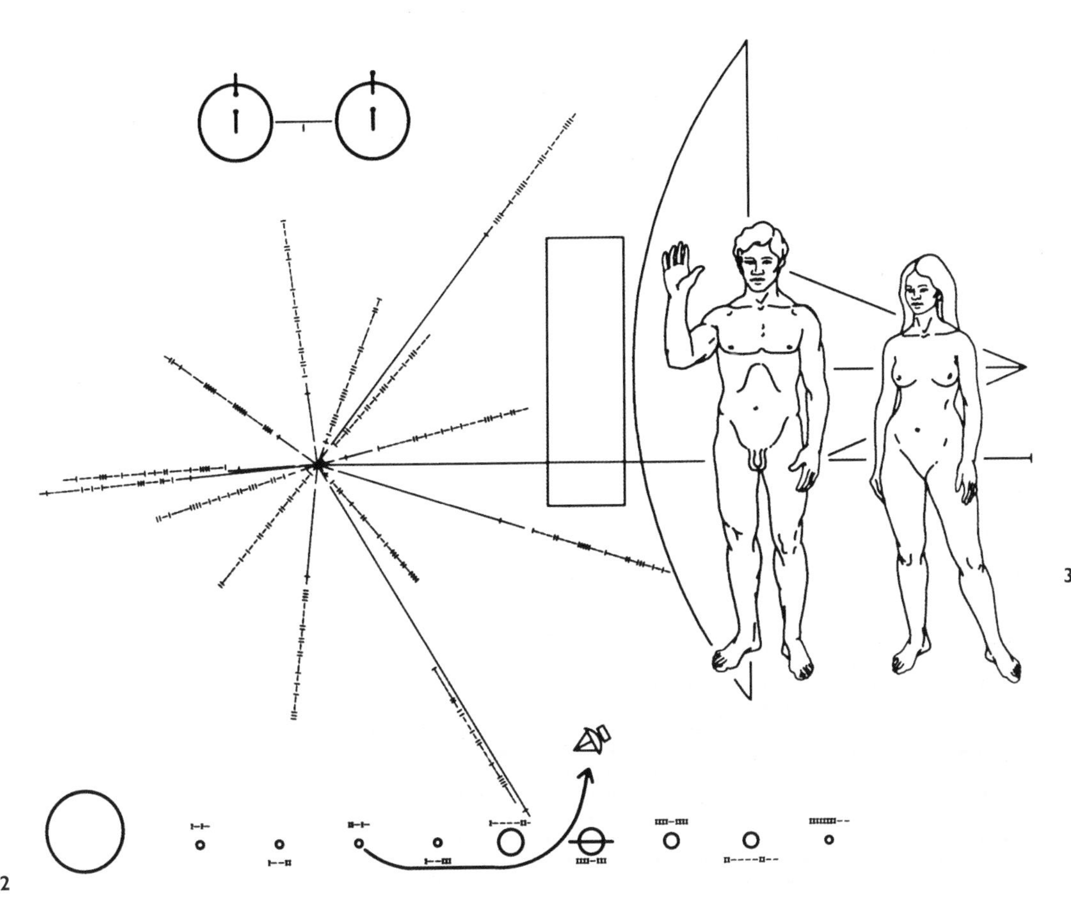

2

3

1. Frank Drake, with his eponymous equation.

2 and 3. The two Pioneer space probes which flew past Jupiter on their way out of the Solar System, each carried a plaque with scientific information and representations of a man and a woman.

opportunity provided seemed too good to miss.

The spacecraft Pioneer 10 was the first human artefact to leave the Solar System. Launched in 1972, Pioneer 10 and its twin probe Pioneer 11 each flew past Jupiter and then followed slightly different trajectories to provide, between them, our first close-up views of all the gas giant planets. The landmark date was 13 June 1983, when Pioneer 10 crossed the orbit of Neptune and officially left the Solar System (Pluto, which is usually the furthest planet from the Sun, occasionally dives just inside Neptune's orbit, and did so between 1979 and 1999). It became the first message in a bottle from the human race to be launched into the ocean of interstellar space.

Each of the Pioneer 10 and 11 spacecraft carried an identical plaque, designed by Carl Sagan and Frank Drake, as a 'hello' to any intelligent beings that might find it. The plaque includes a representation of the spaceprobe itself, the Sun and Solar System, and a 'map' indicating the location of the Sun relative to some of the pulsars. It also includes a rather stylized drawing of a naked man and woman, which drew an angry response from some American citizens concerned that NASA was polluting the Universe with pornography.

When the next missions to Jupiter and beyond were launched in 1977, they carried a more sophisticated message, in the form of a video disc. Each of the two Voyager craft carried a copy of the disc, sealed in a container engraved with scientific information including instructions on how to play the disc. The disc itself contained images of the Earth, sounds ranging from whale song to Chuck Berry, scientific information, and messages from UN Secretary General Kurt Waldheim and US President Jimmy Carter. In truth, any alien who finds the disc is unlikely to learn any more from its contents than the fact that we exist. All subsequent missions to the outer planets have gone into orbit around them, and only Pioneers 10 and 11 and Voyagers 1 and 2 have left the Solar System (▷p. 156).

3.3 LISTENING TO THE STARS

From the time when the technology to transmit and receive radio signals was developed (more or less at the end of the nineteenth century) to the time when live TV transmissions were made from the first astronauts to reach the Moon to anyone on Earth who had the right equipment to receive them was less than 70 years – less than a single human lifetime.

Once a technological civilization invents, or discovers, radio, it is likely to move on rapidly to cheap, powerful radio transmissions that are capable of being detected by comparable civilizations at distances of tens of parsecs. To many people this suggests that it makes more sense for us to concentrate on developing technology to listen for alien broadcasts rather than trying to signal to them ourselves. Such efforts to detect extraterrestrial intelligence are already underway.

Previous page. Optical image of comet Hale-Bopp showing both its gas and dust tails. The gas or 'ion' tail (blue) consists of gas blown away from the comet head by a solar wind.

THE INFINITE RADIO SET

Although radio is likely to be invented when a technological civilization is young, a more advanced civilization will still know about radio waves. They will be aware that this is a good way to communicate with people like us, even if they have found better ways to communicate among themselves. The best way to make contact with a newly emerging technological civilization (which, for all our achievements to date, is what we are) is by radio. This is the reason why the first generation of searches for extraterrestrial intelligence, roughly from the early 1960s to the end of the twentieth century, concentrated on the search for radio signals. But where do you search?

The first thing to decide when carrying out a search for extraterrestrial intelligence by listening out for their radio signals is which radio frequencies to listen at.

The important point about all SETI programmes carried out so far is that they are looking for deliberate signals – beacons of some kind designed to attract attention. Our

1. Radio astronomy observatories such as the Very Large Array link several radio telescopes together to mimic the power of a much larger instrument.

technology is not yet good enough to be able to eavesdrop on communications that are not intended to be picked up by people like us – the alien equivalent of *Monty Python*, perhaps. At first sight the search is a daunting task, because the aliens could be transmitting on any one of an infinite number of radio frequencies. It's no good if our radio receivers are tuned in to the cosmic equivalent of Radio 1 if the aliens are transmitting on Radio 4. We'd never hear them. But nature does impose some limits on the possibilities, and astronomers have also tried to get inside the minds of the aliens to work out logically the frequencies they're most likely broadcast on.

Limiting the Options

The first limitation is imposed by the atmosphere of the Earth. The atmosphere blocks out radio waves with frequencies outside the range from about 1000 megahertz (MHz) to 10,000 MHz. We physically cannot listen outside this range (except with instruments flown in space, above the obscuring layer of atmosphere, and no SETI project has yet gone that far).

The first and most important thing that astronomers learned about the Milky Way using radio telescopes is that it is full of hydrogen gas. This gas radiates radio noise at a frequency of 1420 MHz (equivalent to a wavelength of 21 cm). This enabled radio astronomers to map the spiral arms of our Galaxy by looking at where the hydrogen clouds emitting this radio noise were located.

1. The 43-metre diameter dish antenna of the Green Bank Observatories Radio Telescope.

2. A map of the Milky Way made by radio telescopes studying emission from hydrogen gas.

4

civilization – in some ways, dolphins are intelligent, but they don't have technology.

Counting Earths

The final piece of the puzzle concerns the longevity of such a civilization – its lifetime, *L*, expressed in years. In the 1960s and 1970s, the threat of nuclear war made some pessimists set this figure very low – no more than 100 years. Today, many people are more optimistic, but this remains a big uncertainty. Indeed, in the eyes of the pessimists we have simply replaced the threat of nuclear annihilation by the threat that we will destroy our planet through pollution, bringing an end to our present civilization before the end of the twenty-first century.

When you put everything together, you have an equation for the number of advanced civilizations actively involved in communicating across the Galaxy today.

$$\boldsymbol{N = Rf_s f_p n_e f_i f_i f_c L}$$

This is Drake's equation. If *R* is about 20, and the next six factors multiply up to give a value of about 0.05, which is plausible, what you end up with is the conclusion that *N* is roughly equal to *L*. This means that if civilizations do avoid blowing themselves to bits or polluting their planets to destruction, there may be a very large number of them out there trying to make contact with one another, and with us.

3. Even without warfare, technological civilizations may find ways to destroy themselves and the wildlife around them.

4. Intelligent life could exist deep beneath the clouds of a giant planet like Jupiter without even knowing that the rest of the Universe exists.

TOPIC LINKS

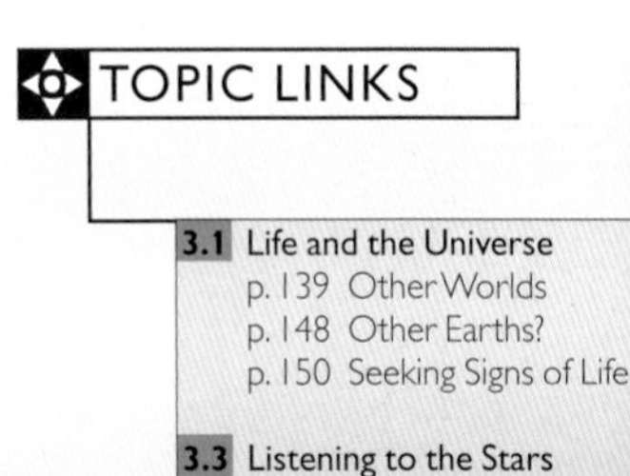

3.1 Life and the Universe
p. 139 Other Worlds
p. 148 Other Earths?
p. 150 Seeking Signs of Life

3.3 Listening to the Stars
p. 172 Life As We Don't Know It

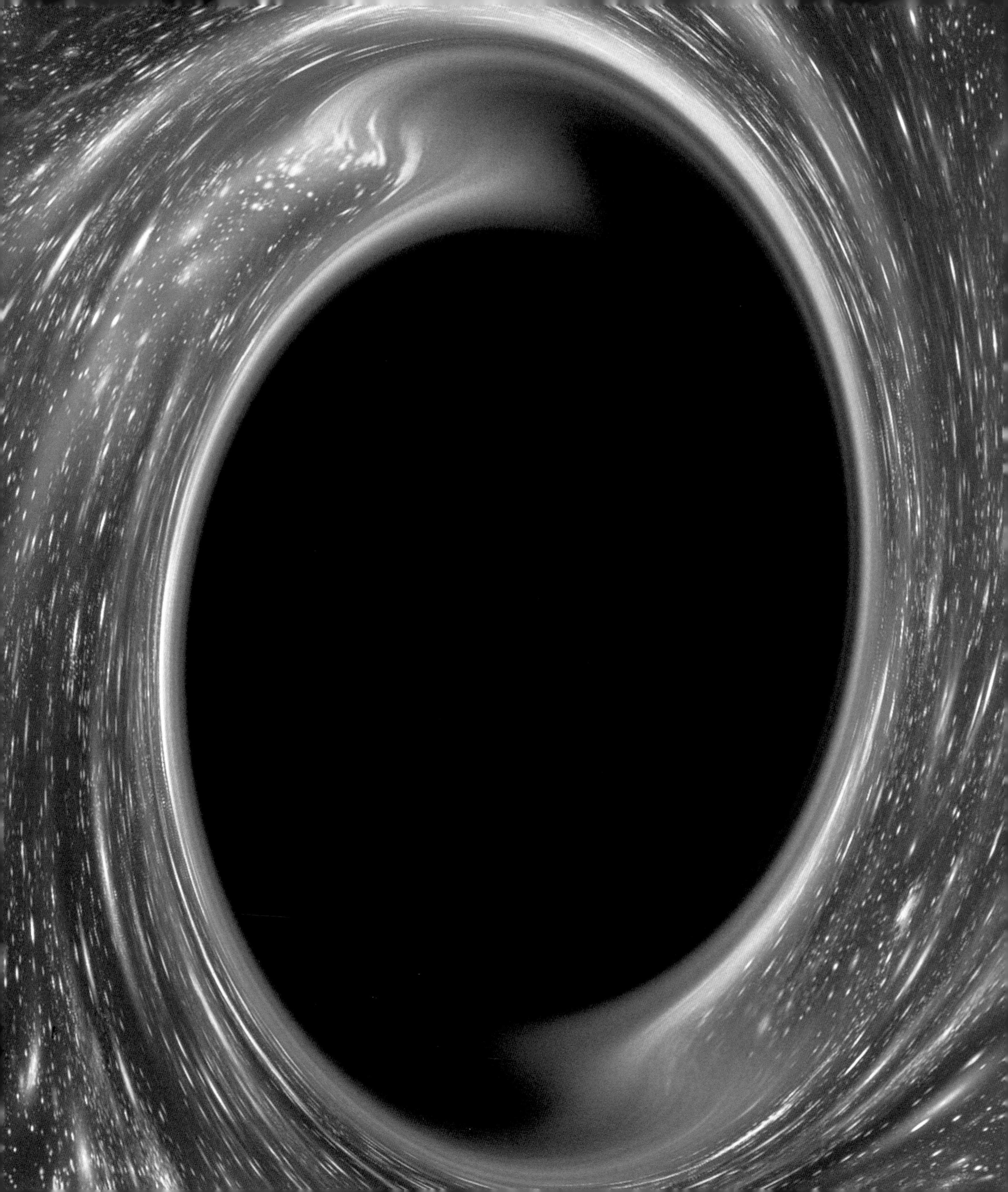

and forms a characteristic pattern of light and shade (an 'interference pattern') on a second screen. This exactly echoes the way ripples on a pond interfere with one another. Similar experiments with beams of electrons show them behaving as waves.

But when experiments are designed to measure the particle properties of electrons or photons, recording their arrival at a target like a stream of little bullets, they turn out to be behaving just like a stream of little bullets. Quantum entities are both particle and wave – a phenomenon called 'wave–particle duality'. It seems that quantum entities travel like waves, but arrive as particles. The equation that describes how they move is called a 'wave function'.

Many Worlds

Wave–particle duality is only the beginning of the story of quantum mysteries. The way a travelling quantum entity 'decides' what kind of particle it is when it is observed is also unlike anything in our everyday experience. As a simple example, think of a single electron travelling through space. When electrons are measured, among other things they each have a property called 'spin'. This is not like the spin of a top, or the Earth on its axis. It can best be thought of as a 'label' on the electron. All that matters is that the measured spin of an atom can only have one of two values, called 'up' and 'down'. A measured electron always has one, and only one, of these spins.

What of a travelling electron? The standard interpretation of quantum theory says that when an electron is on its own it does not have a definite spin, but exists in an indeterminate 50:50 state, a mixture of spin up and spin

THE REALITY OF DUALITY

People sometimes think that wave–particle duality is simply a statistical effect. After all, the waves on the ocean are actually made up of billions of tiny particles – atoms. But quantum wave–particle duality is not like that – it operates at the level of individual entities such as photons and electrons, with the 'waves' having about the same size as the 'particles'. Although there had been indirect evidence of this phenomenon since the 1920s, the reality of duality was brilliantly confirmed in the early 1990s, in an experiment devised by a team of Indian theorists and carried out by a team of Japanese experimental physicists. It took so long to provide this final confirmation, because until then the technology was not up to the task. The experiment involved sending single photons (individual particles of light) through a tiny air gap between two blocks of glass (two prisms), and monitoring their behaviour.

The experiment required great precision, not just in producing single photons but targeting them through the gap between the two prisms, with the size of the gap controlled to within a few tens of a billionth of a metre, about one-tenth of the wavelength of the light involved. Because the gap is so small and light travels as a wave, it could cross the gap, but other tests showed that the photons that arrived on the other side of the gap were particles. Individual photons had been observed behaving as both wave and as particle in the same experiment. Dipankar Home, the leader of the Indian team, summed up the implications. 'Three centuries after Newton,' (left) he said, 'we have to admit that we still cannot answer the question "what is light?"'

1. (opposite) A rainbow is produced when lightwaves are refracted and reflected by raindrops.

1. John Wheeler, who encouraged Hugh Everett's work on the 'Many Worlds' idea.

2. According to quantum theory, the Universe is like a many-branched tree, with each branch corresponding to a different reality.

down. This 'superposition of states', as it is called, applies to all quantum properties. It is only when a measurement is made that there is a 'collapse of the wave function' and the electron chooses (at random) which spin (or whatever) to have. Einstein hated the idea of this random behaviour, and famously declared 'I cannot believe that God plays dice.'

But there is an alternative interpretation. Instead of one electron in a superposition of states, the equations work just as well if there are two electrons, one in each possible state, but in two separate universes. Whichever universe you are in, if you measure the spin you get a definite answer, with (in this case) a 50:50 probability of either choice. For more complicated situations (for example, throwing six dice at once), the odds are more complicated, but there is no collapse of the wave function. On this picture, as Dorothy Parker might have said, 'an electron is an electron is an electron.' The cost is that there is now a separate universe – a separate physical reality – for every possible outcome of every possible quantum measurement. This is where the name 'Many Worlds Interpretation' comes from.

Quantum Physics that Suits Cosmologists

Cosmologists like the idea of many worlds because they have great difficulty with the standard interpretation of quantum mechanics. It is possible to imagine the whole Universe being described mathematically by a single quantum wave function. Such a wave function is sometimes known as the Wheeler–DeWitt equation, after two physicists who studied the problem, but we don't actually know what the equation is, only some of the properties it must have.

The problem cosmologists then have with the standard interpretation is that the Universe is everything there is, so there is nothing outside the Universe to interact with the wave function and make it collapse – it has to exist forever in a superposition of all possible states. This is essentially the same as saying that all possible universes exist side by side, and all instants of time exist (one in front of the other), with nothing really changing. This literally means that the science fiction idea of alternative histories (a world where the South won the American civil war, a world where Nelson Mandela was never imprisoned, and so on) becomes part (a rather small part) of the quantum mechanical description of reality. What are the implications for ideas such as the Big Bang and the expanding Universe?

Quantum Cosmology

There are two kinds of quantum cosmology. The first kind deals with events that occurred very close to the Planck time (time zero) in 'our' Universe (▷ p. 98), and which led to inflation and the Big Bang in which hydrogen and helium were created out of pure energy in a little less than four minutes (▷ p. 107). The second kind of quantum cosmology deals with the implications of the Many Worlds Interpretation and the Wheeler–DeWitt equation. It is much more speculative and exploring these ideas takes us beyond the frontiers of what is known about the way the world works. But it is this kind of exploration that makes the unknown known, and which has already made the first kind of quantum cosmology a respectable part of astronomical thinking.

After his PhD, Hugh Everett worked on classified subjects for the Pentagon and never published another scientific paper.

2

1. Stephen Hawking, a pioneering quantum cosmologist.

2. To a polar bear standing at the North Pole, all directions are south.

1

In 1906, J.J. Thomson received the Nobel Prize for proving that electrons are particles. In 1937, his son, George Thomson, received the Nobel Prize for proving that electrons are waves. Both were right.

The Cosmic Desert

Accepting, for the moment, the everyday idea of time flowing inexorably forward, we can picture the cosmic version of the Many Worlds Interpretation as implying that the Universe was split into many different branches, like a huge tree (sometimes called the 'Multiverse'), by quantum processes operating at the beginning of time, when what is now the observable Universe was the size of a quantum entity. The different branches of the Multiverse would still be in some sense members of the same family, and governed by some common overriding principles (especially the quantum principles which describe the splitting process and allow the universes to keep branching repeatedly as time passes). But there will be an enormous number of different universes (possibly infinitely many) within the Multiverse, and among that infinite array there will be universes with all possible values (and combinations of values) of the fundamental parameters such as omega and lambda, or the Hubble parameter. There could even be

1. Edwin Schrödinger, inventor of the 'cat paradox'.

2. The Schrödinger's cat thought experiment devised by Edwin Schrödinger (opposite) says that a live cat and its ghost can both exist at the same time.

superposition of states until somebody looks into the room and notices what has happened. The cat is both dead and alive at the same time.

Parallel Possibilities

There are several rival interpretations of quantum mechanics which try to avoid this unwelcome state of affairs. The one that many cosmologists like involves parallel worlds. In this picture, the moment the electron is released, the entire world splits into two copies of itself. In one, the electron has spin down and the cat lives. In the other, the electron has spin up and the cat dies. For a human observer in either world, there is still a 50:50 chance of finding a live cat when you look into the room – but neither cat is ever in a superposition of states.

Extending this example, the entire Universe is multiplied into an infinite number of branches, and anything that can possibly happen does happen in one (or more) of the branches of reality. (This experiment really is 'all in the mind'. Nothing like it has ever been tried with a real cat!)

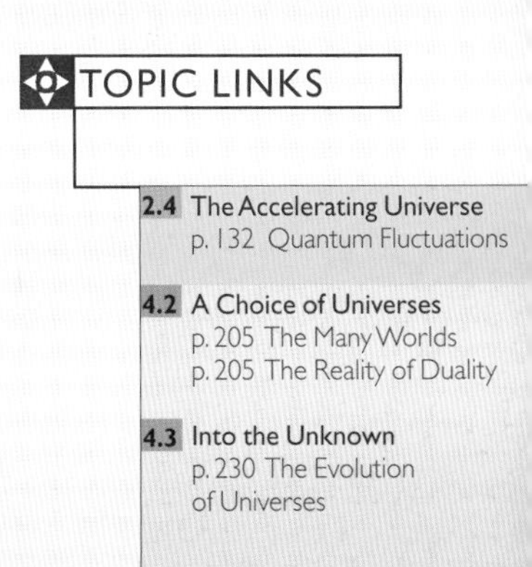

TOPIC LINKS

2.4 The Accelerating Universe
p. 132 Quantum Fluctuations

4.2 A Choice of Universes
p. 205 The Many Worlds
p. 205 The Reality of Duality

4.3 Into the Unknown
p. 230 The Evolution of Universes

4.3 INTO THE UNKNOWN

The idea that alternative worlds might exist somewhere in a quantum Multiverse, combined with the weak anthropic principle, is a neat way of explaining why the Universe we see around us seems to have been designed for life, without invoking a Designer and becoming embroiled in the infinite regress of who designed the Designer, and so on. But all this is rather abstract and philosophical.

The quantum rules do, however, allow for a much more direct relationship between universes, suggesting that they may be physically connected, and that one universe may grow out of another, through a black hole. According to these ideas, Universes reproduce, producing baby universes which grow and have babies in their turn. These ideas come from research at the far frontier of science, and have not yet been tested. But they show where cosmological thinking is taking us in the twenty-first century.

BEYOND THE BLACK HOLE

The equations of the general theory of relativity (a very well established theory that has been tested by many experiments) tell us that the entire contents of a black hole must collapse towards a mathematical point, a singularity (▷ p. 87). The equations of quantum physics (an equally well founded theory) tell us, however, that there is no such thing as a mathematical point, and that nothing can be smaller across than the Planck radius.

Combining these two well-established ideas, physicists conclude that something happens to matter (mass-energy) falling inwards inside a black hole when it reaches the Planck radius. The most likely result is that the material bounces, and expands outwards again. But it doesn't come bursting back out into the universe it fell in from; instead, it is shunted sideways into a new set of dimensions, forming a new expanding universe in its own right.

To put these ideas in perspective, we need to remind ourselves what kind of objects black holes are.

Black Holes Revisited

A black hole is formed by any concentration of matter which has a gravitational field so strong that spacetime is bent right around to form a closed surface. There are two ways this can happen. If any lump of matter is squeezed into a ball so that, although its mass stays the same, its density increases, at some critical density it will become a black hole. This is the kind of black hole left behind by some supernovae. Alternatively, if you keep the density constant and keep adding more mass, at some critical mass it will become a

1

1. (opposite) Close-up view of the jet of gas being emitted from the black hole at the centre M87.

black hole. This is the kind of black hole that powers a quasar.

The crucial radius at which an object becomes a black hole is called the Schwarzschild radius, in honour of the German scientist who first realized that the equations of the general theory of relativity predicted the existence of black holes. One way of thinking of the Schwarzschild radius is that the escape velocity from a surface with the Schwarzschild radius is the speed of light. Nothing can escape from inside the Schwarzschild surface, because within it the escape velocity exceeds the speed of light.

To see how the different ways of making black holes work, think of the Sun. If the Sun were squeezed within a ball just 2.9 km across (the Schwarzschild radius for an object with one solar mass), it would become a superdense black hole. But if you could add more matter to the Sun without it collapsing (think of a bag containing a lot of marbles, each representing a star like the Sun), it would become a black hole when it had the mass of a few million Suns, and about the radius of our Solar System – the Schwarzschild radius for an object with a few million solar masses. The density of matter needed to make such a supermassive black hole would only be a little more than the density of water – but the matter would then fall inward towards the centre and be crushed out of existence at the Planck radius.

Einstein's Wormholes

Although black holes were only given their modern name in 1967 (by John Wheeler),

THE MAN WHO INVENTED BLACK HOLES

The first person to suggest the existence of black holes was an eighteenth-century English parson, the Reverend John Michell. But he wasn't just a country parson. Michell, who was born in 1724, became one of the leading scientists of his day before he took Holy Orders, and made his reputation from a study of the disastrous earthquake that struck Lisbon in 1755 (left). He was elected a Fellow of the Royal Society in 1760, and Professor of Geology at the University of Cambridge in 1762. In 1764, he left the university to become the rector of a parish in Yorkshire, but he maintained a keen interest in science and published several important astronomical papers. Among other achievements, he was one of the first people to publish a reasonably accurate estimate of the distance to a star. Using an argument based on its apparent brightness, he calculated that the star Vega is 460,000 times further away from us than the Sun is. In 1783, in a paper read to the Royal Society by his friend Henry Cavendish, Michell pointed out that there could be 'dark stars' in the Universe, so big that light could not escape from them:

'If there should really exist in nature any bodies whose density is not less than that of the Sun, and whose diameters are more than 500 times the diameter of the Sun, since their light could not arrive at us...we could have no information from sight; yet, if any other luminiferous bodies should happen to revolve around them we might still perhaps from the motions of these revolving bodies infer the existence of the central ones.'

These are exactly the kind of black holes now thought to be associated with quasars, and we do indeed infer their existence from the way bright stuff orbits around them.

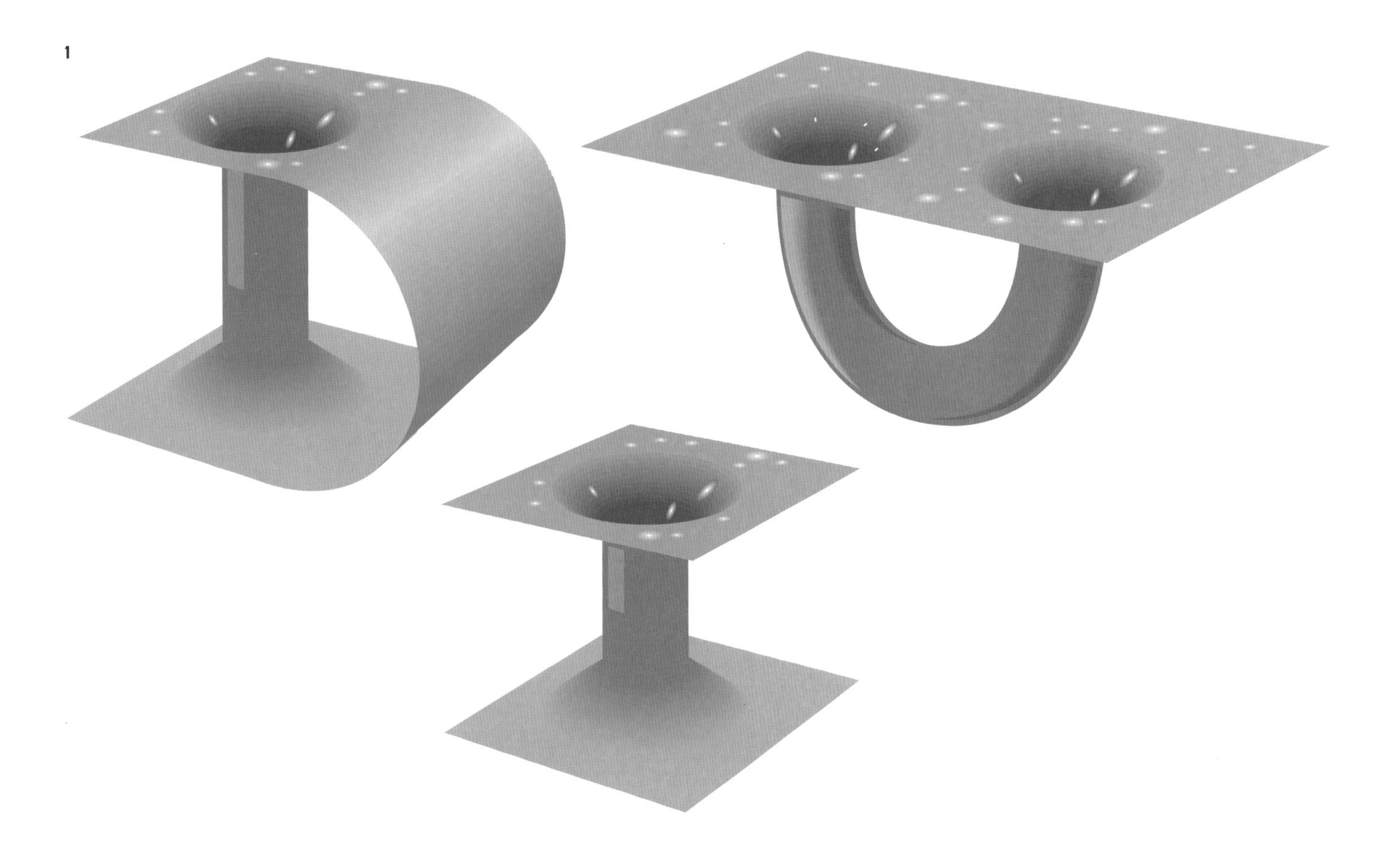

1. 'Wormholes' may join different regions of a single universe, or even separate universes.

Karl Schwarzschild predicted their existence in 1916, and the mathematical description of black holes was studied by Albert Einstein himself in the 1930s. Working with Nathan Rosen, at Princeton University, Einstein discovered that the solutions to the equations of the general theory discovered by Schwarzschild actually described not a single 'hole in space' but a pair of holes connected by a tunnel linking two separate regions of spacetime. This tunnel became known as an Einstein–Rosen bridge, but more recently physicists have taken to calling them 'wormholes'.

An Einstein–Rosen bridge, or wormhole, joins two black holes in different locations in spacetime. Originally, people thought that if such entities were real they could link different parts of our Universe, like a cosmic subway. This is still true; but because space*time* is involved not just space, a wormhole might also, in principle, link two different times in our Universe – it could be a kind of time machine. And now cosmologists speculate that if other universes exist, then wormholes may provide links between different universes.

Quantum Foam

Don't get too excited, however, about the possibility of travelling through a wormhole to some other place, or time, or some other universe. It would be extremely difficult (for all practical purposes, it may be impossible) to build a large wormhole through which people could travel. Natural wormholes, if they occur at all, exist on the scale of the Planck length, whatever the size of the Schwarzschild radius of the black hole that acts as the gateway to a particular cosmic subway.

Physicists who study the mathematics of such entities are still intrigued by wormholes, however, because quantum processes operating on the Planck scale may produce vast numbers of tiny (sub-sub-microscopic) wormholes, and these could provide the structure of spacetime itself. The quantum wormholes would be like the strands of a carpet, woven together to make the seemingly solid structure of spacetime (the carpet) itself. And they could be the seeds of baby universes.

1

1, 2 and 3 (opposite): The surface of the sea looks less and less smooth the closer you get to it.

2

HOW TO MAKE A UNIVERSE

The idea that spacetime may be woven together out of quantum-scale wormholes is related to another of John Wheeler's suggestions, which portrays spacetime as a 'foam' of quantum entities, popping in and out of existence at the scale of the Planck length. This could include black hole pairs and the wormholes that connect them. An analogy Wheeler makes is with the appearance of the surface of the sea. From a high-flying aircraft, the sea looks smooth and flat. Closer up, it looks rough, and closer still the surface dissolves into a constantly changing foam of bubbles and tiny waves. Spacetime may only seem smooth to us because we are so much larger than the Planck scale – remember that the Planck length, 10^{-33} cm, is 10^{-20} times the size of a proton.

Virtual Reality

Wheeler's idea is based on the prediction from quantum mechanics that quantum entities such as pairs of particles can appear out of nothing at all for a short time ('virtual pairs'). We met this idea in Chapter 2, in the

1. A pair of interacting galaxies known as the antennae.

2. A gamma ray burster pictured by the Hubble Space Telescope.

1. A jet of material in the galaxy M87 being shot out by a supermassive black hole at the heart of the galaxy.

in this way. Little packets of pure energy (bubbles with a radius roughly equal to the Planck length) can also appear out of nothing at all, provided that they disappear within the time allowed by Heisenberg's uncertainty relation. But that time can be very long indeed.

Remember that the less energy a quantum fluctuation contains, the longer it can live. The energy of a little packet of mass-energy actually comes in two forms – from its mass, but also from its gravity. Curiously, the energy of a gravitational field is actually negative. This means that for a bubble with just the right amount of mass-energy the two effects cancel out, and the overall energy of the bubble is zero. In that case, it could last forever. 'Just right' means that the bubble contains just enough mass to make it flat, on the edge of being a black hole. If our Universe is flat, it also contains zero energy overall.

As long ago as 1973, Edward Tryon, of City University in New York, pointed out that as far as quantum physics was concerned the entire Universe could be nothing more or less than a quantum fluctuation of the vacuum.

Blowing Bubbles

The big snag with Tryon's idea was that quantum fluctuations occur on the Planck scale. There was, indeed, nothing in the

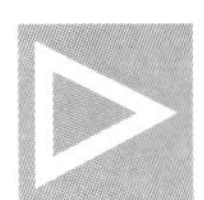

EINSTEIN'S SURPRISE

The idea that the negativity of gravity can precisely cancel out the mass-energy $E=mc^2$ in a lump of matter (or the Universe) is so surprising that even after you have had it explained to you, or studied the equations for yourself, it is hard to believe. If you feel that way, you are in very good company.

The first person to realize this possibility was the physicist Pascual Jordan, working in the United States in the 1940s. At that time, Albert Einstein (right) worked as a consultant for the US Navy, assessing schemes for new weapons sent to the military by well-meaning civilians. Einstein was very good at this – in the first decade of the twentieth century he had worked as a Patent Officer in Switzerland, so he was used to finding the flaws in inventions. Every fortnight or so, George Gamow, who was based in Washington DC and also involved in war work, would take the latest batch of ideas up to Einstein in Princeton.

On one of these visits, as Gamow later recalled in his autobiography *My World Line*, the two physicists were walking from Einstein's home to the Institute for Advanced Study, where Einstein worked, when Gamow casually mentioned that Jordan had told him that a star could be made out of nothing at all, because at the point of zero volume its negative gravitational energy would precisely cancel out its positive mass-energy. 'Einstein stopped in his tracks,' Gamow tells us, 'and, since we were crossing a street, several cars had to stop to avoid running us down.'

quantum rules to forbid a bubble containing as much mass-energy as our entire Universe to appear on the Planck scale. But it would have an enormous gravitational pull, which would crush it out of existence immediately. Tryon's idea lay dormant until the 1980s, when inflation theory provided the mechanism which can take a superdense quantum seed and blow it up to the size of an observable object in our world. The enormous push given by inflation acts like antigravity, flattens spacetime, and prevents the baby universe crushing itself out of existence as soon as it forms.

There is no reason, however, to stop at one universe. In this picture, universes of all sizes should be produced by quantum fluctuations, some with just enough strength to expand a little way, others where the Λ term is powerful enough to keep the universe expanding forever, and with all possible intermediate values. Each universe is like another bubble in the foam of spacetime, expanding in its own way with no contact with its neighbours, and we are back to the weak anthropic principle to explain why the Universe we live in is as it is.

But Tryon's ideas, inflation, and the mathematics of black holes can be combined to give a more intriguing insight into how we got here.

Black Holes and Baby Universes

The seed of a universe like our own could come into existence as a concentration of mass-energy containing all the mass-energy of the observable Universe in a volume with the Planck radius, 'born' at an age of 10^{-43} seconds, the Planck time (▷p. 98). Inflation could drive the expansion of such a baby universe up to a large size, and this could be happening anywhere in the vacuum of space around us (or, indeed, in the space between the atoms of the page you are now reading). If so, these other universes wouldn't explode outwards in your face, filling up our own region of spacetime, but would exist in their own sets of dimensions, all of them at right angles to the dimensions of our spacetime. But it would also be possible, in principle, to make a universe, deliberately.

Universes on Demand

Several physicists, including Alan Guth, one of the pioneers of inflation theory, have explored this possibility mathematically. One of the key points in their argument is that even in order to make as large a Universe as the one we live in, you do not need a lot of mass-energy to start with. Because of the negativity of gravity, nature can make universes out of nothing at all. We cannot quite do that, because we would have to put energy in to trigger the process of inflation. But the amount of energy required is surprisingly small, compared even with the output of a star like the Sun.

The requirement to kick-start inflation is a temperature equivalent to about 10^{24} K and a very high density. The energy required to produce these conditions is no more than could be produced by a few hydrogen bombs; the problem is confining that energy, even if only for a split-second, in a tiny volume of space, about the size of an atom. If this could be done, the equations studied by Guth and others show that under some circumstances inflation will occur in the compressed region.

1. The technology of the hydrogen bomb almost gives us the ability to make black holes – and perhaps baby universes.